AF775571

Vector-Borne Diseases - Recent Developments in Epidemiology and Control

Edited by David Claborn, Sujit Bhattacharya and Syamal Roy

Published in London, United Kingdom

IntechOpen

Supporting open minds since 2005

Vector-Borne Diseases - Recent Developments in Epidemiology and Control
http://dx.doi.org/10.5772/intechopen.83110
Edited by David Claborn, Sujit Bhattacharya and Syamal Roy

Contributors
Arijit Chakraborty, Malini Sen, Shreyasi Maity, Pijush K Das, Anindita Bhattacharya, Arunima Biswas, Susanta Ghosh, Chaitali Ghosh, Etienne Waleckx, Anette Hernandez-Andrade, Joel Moo-Millan, Nohemi Cigarroa-Toledo, Angel Ramos-Ligonio, Claudia Herrera, Bruno Bucheton, Jean-Mathieu Bart, Anne-Laure Bañuls, David Roiz, Denis Sereno, Carlos N. Ibarra-Cerdeña, Carlos Machain-Williams, Julián Garcia-Rejon, Sébastien Gourbière, Christian Barnabé, Jenny Telleria, Bruno Oury, Simone Frédérique Brenière, Frederic Simard, Miguel Rosado, Philippe Solano, Eric Dumonteil, Vincent Jamonneau, Christophe Paupy, Rafaela Bruno, Luciana Araripe, Luana Farnesi, Melaku Wale, Alemayehu Abate, Marta Giovanetti, Vagner Fonseca, Joilson Xavier, Ana Maria Bispo De Filippis, Luiz Carlos Junior Alcantara, Tulio De Oliveira, San Emmanuel James, Lisette Tongue Kohagne, Arouna Njayou Ngapagna, Pravin Kendrekar, Rajendra Pawar, Sam Mashele, Sunil Tekale, David Claborn

© The Editor(s) and the Author(s) 2020
The rights of the editor(s) and the author(s) have been asserted in accordance with the Copyright, Designs and Patents Act 1988. All rights to the book as a whole are reserved by INTECHOPEN LIMITED. The book as a whole (compilation) cannot be reproduced, distributed or used for commercial or non-commercial purposes without INTECHOPEN LIMITED's written permission. Enquiries concerning the use of the book should be directed to INTECHOPEN LIMITED rights and permissions department (permissions@intechopen.com).
Violations are liable to prosecution under the governing Copyright Law.

Individual chapters of this publication are distributed under the terms of the Creative Commons Attribution 3.0 Unported License which permits commercial use, distribution and reproduction of the individual chapters, provided the original author(s) and source publication are appropriately acknowledged. If so indicated, certain images may not be included under the Creative Commons license. In such cases users will need to obtain permission from the license holder to reproduce the material. More details and guidelines concerning content reuse and adaptation can be found at http://www.intechopen.com/copyright-policy.html.

Notice
Statements and opinions expressed in the chapters are these of the individual contributors and not necessarily those of the editors or publisher. No responsibility is accepted for the accuracy of information contained in the published chapters. The publisher assumes no responsibility for any damage or injury to persons or property arising out of the use of any materials, instructions, methods or ideas contained in the book.

First published in London, United Kingdom, 2020 by IntechOpen
IntechOpen is the global imprint of INTECHOPEN LIMITED, registered in England and Wales, registration number: 11086078, 7th floor, 10 Lower Thames Street, London, EC3R 6AF, United Kingdom
Printed in Croatia

British Library Cataloguing-in-Publication Data
A catalogue record for this book is available from the British Library

Additional hard and PDF copies can be obtained from orders@intechopen.com

Vector-Borne Diseases - Recent Developments in Epidemiology and Control
Edited by David Claborn, Sujit Bhattacharya and Syamal Roy
p. cm.
Print ISBN 978-1-83880-021-5
Online ISBN 978-1-83880-022-2
eBook (PDF) ISBN 978-1-83880-038-3

For EU product safety concerns:
IN TECH d.o.o., Prolaz Marije Krucifikse Kozulić 3, 51000 Rijeka, Croatia, info@intechopen.com or visit our website at intechopen.com.

We are IntechOpen, the world's leading publisher of Open Access books

Built by scientists, for scientists

4,800+
Open access books available

123,000+
International authors and editors

140M+
Downloads

151
Countries delivered to

Our authors are among the
Top 1%
most cited scientists

12.2%
Contributors from top 500 universities

BOOK CITATION INDEX
CLARIVATE ANALYTICS
INDEXED

WEB OF SCIENCE™

Selection of our books indexed in the Book Citation Index
in Web of Science™ Core Collection (BKCI)

Interested in publishing with us?
Contact book.department@intechopen.com

Numbers displayed above are based on latest data collected.
For more information visit www.intechopen.com

Meet the editors

David M. Claborn is Director of the Master of Public Health Program at Missouri State University, where he teaches courses in international health, environmental health, public health preparedness, and homeland security. He obtained a DrPH in 2001 from the Uniformed Services University of Health Sciences in Bethesda, Maryland. His dissertation dealt with the re-emergence of malaria in South Korea. Dr. Claborn also has a MS in Entomology from Texas Tech University. Prior to starting his academic career in 2008, he was a career naval officer working as a medical entomologist. He retired from the Navy's Medical Service Corps in 2008 at the rank of Commander.

Dr Sujit Kumar Bhattacharya was born on 15 January 1950 and obtained his MD (Internal Medicine) degree in 1979 from Calcutta University. He joined the National Institute of Cholera and Enteric Diseases, Kolkata and became Director of the institute in 1994. He was then elevated to Additional Director General of the Indian Council of Medical Research, New Delhi. He joined the South East Asia Regional Office, New Delhi in 2009 and retired in 2011. He also served as Director of the Rajendra Medical Research Institute of Medical Sciences, Patna, and Virology Research Centre, Kolkata. He is a fellow of three prestigious academies in India. He served in the Scientific Advisory Group of Tropical Diseases Research (TDR), Geneva. He attended many national and international conferences. He has to his credit about 500 publications. He received the Ranbaxy Science Foundation award for excellence in research in clinical science. He is now involved in clinical medical patient care.

Dr Syamal Roy obtained his PhD in Biochemistry from the University of Calcutta, India and received his post-doctoral training in Cellular Immunology at The Massachusetts Institute of Technology, Cambridge, USA. He has been working at the Indian Institute of Chemical Biology, Calcutta on his independent research on the immuno-biology of leishmaniasis. He has made substantial contribution in this area of research and currently continues to work in the same area.

Contents

Preface

Vector-Borne Diseases - Recent Developments in Epidemiology and Control reviews many new developments in the study of diseases spread by arthropod vectors. Any group of writers that hopes to adequately address such a topic must represent a diverse group from a geographically dispersed area; the scientists and health workers who wrote this book are such a group. They address the impact of vector-borne diseases for humans and livestock in Africa, the effects of insecticide on entomological inoculation rates in Ethiopia, the effects of viral infection on the primary vector of Zika and dengue in South America, and the potential of certain chemical structures in the treatment of disease. The authors hail from Africa, India, North America and other places, so this book includes a variety of cultures and writing styles, just like the vulnerable populations that suffer from vector-borne infections. This broad presentation of expertise seems overwhelming, yet the continuously evolving disease threat can only be addressed with rapid development of technology and effective sharing of knowledge. The emergence and re-emergence of vector-borne diseases illustrates the need for such knowledge.

In recent years, Zika appears to have taken on a greater measure of pathology and then spread out of its usual confines in Africa to present a disease threat through much of the tropical and some of the temperate parts of the world. Previously thought to be a relatively mild, self-limiting illness, Zika is now known to have devastating effects on children. This example demonstrates the ever-changing threat due to vector-borne diseases, though it may seem like a distant threat to some observers. Such is not the case for other infections. In recent years, two tick-borne viruses capable of causing severe illness, even death, have emerged in the very central part of the United States. The emergence of these tick-borne viruses in the heart of a developed country demonstrates how little is really known about these disease agents. The reality is that there may be hundreds of vector-borne viral diseases still to be detected and described to the medical community.

This book is just a small attempt to share some of the important work going on in the study of vector-borne diseases. It represents a broad slice of research developments, some of which may contribute to new treatments or new methods of disease control. Perhaps ready access to information contained here will contribute to improved care or disease prevention for someone in the world. That is the hope of the authors of the chapters in this book.

David Claborn, DrPH
Master of Public Health Program,
Missouri State University,
Springfield, Missouri, USA

Sujit Bhattacharya
University of Calcutta,
India

Syamal Roy
National Institute of Pharmaceutical Education and Research,
India

Section 1

Introduction to Vector-Borne Disease Risk

Chapter 1

Introductory Chapter: Vector-Borne Diseases

David Claborn

1. Introduction

In the waning years of the nineteenth century, Theobald Smith convincingly proved that Texas cattle fever was caused by a protozoan parasite and, perhaps more importantly, was transmitted to cattle by a tick. This was the first definitive proof of vector-borne transmission by an arthropod [1]. Within a few years, scientists demonstrated transmission of human disease agents by vectors for a variety of diseases, from filariasis to malaria. Since then, vector-borne transmission has proven to be a major means of disease for dozens of diseases, some of great public health importance. **Table 1** provides a partial list of vector-borne diseases along with the disease agents and vectors for each. Malaria is probably the vector-borne disease with the largest impact on human health today, though historically typhus, bubonic plague, and yellow fever have been extremely important [2].

Despite extraordinary efforts to control or eliminate vector-borne diseases, they persist. Estimates of mortality due to malaria alone exceed 400,000 per year [3]. For the millions of survivors of malaria infection, the costs of disease and disability are enormous. And this reflects a recent improvement over previous years. Great effort has gone into reducing the incidence of malaria, reflecting the dedication of governments, nongovernmental organizations, charitable agencies, scientists, and medical workers. Despite significant success with reducing the rates of diseases like malaria, typhus, and yellow fever, vector-borne diseases persist. Of the 20 illnesses listed as neglected tropical diseases by the World Health Organization in 2020, 6 are primarily transmitted by vectors: American trypanosomiasis, African trypanosomiasis, dengue (and chikungunya), leishmaniasis, lymphatic filariasis, and onchocerciasis. Some of the others, like trachoma, may also exhibit a vector-borne element through the mechanical transmission of disease agents by filth flies. The vectors represent a wide spectrum of arthropod species, from the ticks and mites of *Arachnida* to the mosquitoes, true bugs, and lice of *Insecta*. The disease transmission cycles are as diverse as the range of vectors.

Although the disease transmission cycles of vector-borne diseases can be very complex, they do provide a unique opportunity for prevention or control. For non-vector-borne diseases, prevention can take a variety of forms including immunizations, sanitation, infection control, chemoprophylaxis, curative medicine, and others. For vector-borne disease, however, there is often the possibility of vector control as a means of interrupting the disease cycle. Some vector-borne diseases, such as yellow fever and Japanese encephalitis, can be effectively prevented by immunization, though maintenance of a cold chain to ensure vaccine viability may be necessary and can be difficult to maintain in remote environments. Other diseases such as malaria can be prevented through prophylactic use of drugs. But for many vector-borne diseases, there are no effective vaccines or chemoprophylactic measures. Also, there is no specific treatment for many of the diseases. For such

IntechOpen

Disease	Agent	Vector
Babesiosis	Protozoan	Tick
Bartonellosis	Bacteria	Sand fly
Chikungunya	Virus	Mosquito
Crimean-Congo hemorrhagic fever	Virus	Tick
Dengue fever	Virus	Mosquito
Ehrlichiosis	Intracellular bacterium	Tick
Japanese encephalitis	Virus	Mosquito
Leishmaniasis	Protozoan	Sand fly
Louse-borne relapsing fever	Bacterium	Louse
Lyme disease	Bacterium	Tick
Lymphatic filariasis	Filarial worm	Mosquito
Malaria	Protozoan	Mosquito
Onchocerciasis	Filarial worm	Black fly
Plague	Bacterium	Flea
Q fever	Intracellular bacterium	Tick
Relapsing fever	Bacterium	Tick
Rickettsiosis	Intracellular bacterium	Tick, mite
Rift Valley fever	Virus	Mosquito
Rocky Mountain Spotted fever	Intracellular Bacterium	Tick
Sand Fly fever	Virus	Sand fly
Tick-borne encephalitis	Virus	Tick
Tularemia	Bacterium	Tick, deer fly
Typhus	Bacterium	Lice

[1]*This is a partial list, especially for the tick-borne viruses. Mechanical transmission by filth flies and cockroaches is not considered here, though several diseases such as shigellosis probably have vector-borne components of their transmission.*

Table 1.
Important vector-borne diseases, disease agents, and vectors.[1]

illnesses, vector control or protection of humans from exposure to the vector may be the most important means of risk reduction. Such is the case for diseases like dengue, chikungunya, and Zika. Though the vector provides another vulnerability to the disease transmission cycle, it also provides a mechanism for spreading the diseases. The resurgence of vector population like that of *Aedes aegypti* in Latin America has been linked to the resurgence of dengue [2]. Likewise, expansion of the range of invasive species like *Aedes albopictus* in the USA can present an increased risk of disease transmission in affected regions.

The current status of information regarding vector-borne disease prevention, control, and treatment demonstrates the need for more research and dissemination of the knowledge gathered from laboratory and field studies alike. Developments in molecular biology, genomics, pharmacology, field biology, and virology provide great potential for improvements in control of vector-borne disease. At the same time, it is necessary to acknowledge the continuing utility of older, field-proven methods such as interior residual sprays, which continue to provide effective and inexpensive disease control for millions of people [4]. The wide variety of vector-borne diseases along with the fact that they can be attacked through a wide variety of methods makes the study of such diseases far ranging in subject and perspective.

Conflict of interest

The author declares no conflict of interest.

Author details

David Claborn
Master of Public Health Program, Missouri State University, Springfield, Missouri, USA

*Address all correspondence to: davidclaborn@missouristate.edu

IntechOpen

© 2020 The Author(s). Licensee IntechOpen. This chapter is distributed under the terms of the Creative Commons Attribution License (http://creativecommons.org/licenses/by/3.0), which permits unrestricted use, distribution, and reproduction in any medium, provided the original work is properly cited.

References

[1] Smith T, Kilborne FL. Investigations into the nature , causation, and prevention of Texas or southern cattle fever. Bulletin U.S. Dept of Agriculture Bureau of Animal Industry. 1893:1

[2] Wilson AL, Courtenay O, Kelly-Hope LA, Scott TW, Takken W, Torr SJ, et al. The importance of vector control and elimination of vector-borne diseases. PLoS Neglected Tropical Diseases. 2020;**14**(1):000007831. DOI: 10.1371/journal.pntd.0007831

[3] World Malaria Report. Geneva: World Health Organization, 2019. License CC BY-NC-SA 3.0 IGO; 2019

[4] Townson H, Nathan MB, Zaim M, Guillet P, Manga L, Bos R, et al. Exploiting the potential of vector control for disease prevention. Bulletin of the World Health Organization. 2005;**85**:942-947

Chapter 2

Emerging Vector-Borne Diseases in Central Africa: A Threat to Animal Production and Human Health

Lisette Kohagne Tongue and Arouna Njayou Ngapagna

Abstract

Although the potential for livestock production is high in Central Africa, it is not an important economic activity because of disease constraints, primarily trypanosomiasis transmitted by tsetse flies. Recently, a growing number of vector-borne diseases have also emerged in that region. Indeed, there is a progressive expansion of trypanosomiasis in known tsetse-free areas in the Far North of Cameroon, mechanically transmitted by Tabanidae. In the beginning of year 2019, there was an epidemic of African horse sickness (AHS) in Cameroon for the first time. In the meantime, AHS was also declared in Chad and reported in Nigeria. Besides, new cases of Rift Valley fever (RVF) are regularly detected in both Cameroon and Chad. The relative significance of most vector-borne diseases (VBDs) in livestock is difficult to quantify, because there is no study on their socioeconomic impact. But, certain VBDs have significant impact on food production, and others such as RVF can be transmitted to humans. Impact of VBDs on human health, animal health and trade, as well as the transboundary nature of these diseases means there is a need for regional coordination and cooperation to address challenges. This can be successfully achieved with One Health approach.

Keywords: trypanosomiasis, African horse sickness, Rift Valley fever, Cameroon, Chad

1. Introduction

sub-Saharan Africa has been classified into five agroecological zones (AEZs): arid, semiarid, subhumid, humid, and highlands [1]. AEZs are one of the most important determinants of the characteristics of livestock production systems, in terms of species, breed, stocking capacity, disease pressure, and individual productivity. In sub-Saharan Africa, the rural population lives mainly from agriculture and livestock. Ruminant livestock are found mainly in arid and semiarid zones in the following numerical order: goats, sheep, and cattle in arid zone and cattle, goats, and sheep in semiarid and subhumid zones [2]. Livestock production in this zone is usually a component of mixed smallholder crop-livestock systems. Pastoral system comprises 21 percent of total cattle numbers. About 30 percent are kept in the mixed semiarid system, 21.7 percent in the mixed subhumid, and only 3.6 percent in the mixed humid system [3].

IntechOpen

Traditional ruminant production systems in sub-Saharan Africa are generally subdivided into two broad categories: grassland-based systems and mixed systems. In West and Central Africa, the main production system is mixed systems, which means the farming system is (i) based on livestock but practiced in proximity to, or perhaps in functional association with, other farming systems based on cropping such as pastoral systems in arable areas [4] and is (ii) characterized by a long tradition of seasonal penetration into the more humid areas, with southward movements during dry season and northward movements during rainy season [5]. Although the potential for livestock production is high in humid zone that occupies 112 million ha in Central Africa [6], it is not an important economic activity there because of disease constraints, primarily trypanosomiasis transmitted by tsetse flies [7]. Trypanosomiasis is not the only vector-borne disease that affects animal production in Africa. Indigenous African transboundary diseases such as African horse sickness (AHS), bluetongue, and Rift Valley fever (RVF) are diseases that can cause high mortality among animal population and decrease animal production.

A vector-borne disease can simply be defined as a disease transmitted by a living being, usually an arthropod vector, to a vertebrate host depending on a balance between the vector, the pathogen transmitted, and the host. Disease occurs in an area when all three components are in place, vector's density being determined by climatic conditions. As such, incidence of vector-borne disease is closely related to the presence of the vector. Climate change leads to changes in the geographical distribution of the vector which often has an influence on the epidemiology of the disease. In its extreme form favored by the introduction of a new pathogen, it can lead to the emergence or re-emergence of a disease.

This review is to highlight emerging vector-borne diseases in the Central Africa region and possible ways of control.

2. Vectors and transmission of infectious diseases

Vectors are living organisms, mainly arthropods that can transmit pathogens during their blood meal between vertebrate hosts including humans. Mouth parts of bloodsucking insects are adapted for piercing the skin of animal hosts and sucking their blood. Not all vectors are strictly hematophagous. For some species such as mosquitoes, blood feeding is needed for egg production. Thus, only females are bloodsuckers, while other species such as tsetse flies need blood meal for their survival, and both males and females are bloodsuckers. Arthropods can feed on various orders of mammals (although some have preferences) and contribute to spread pathogens between taxonomic groups.

Ticks (Ixodidae or "hard ticks") are main vectors of animal diseases found in countries with warm and humid climates [8, 9]. Their preferential habitats are forests, savannahs, grasslands, and scrublands, in which they find suitable environmental conditions for their survival [10]. However, the majority of arthropod vectors belong to four orders of hematophagous insects, namely, Phthiraptera (lice), Siphonaptera (fleas), Diptera (flies), and Hemiptera (true bugs). Fleas are vectors of various pathogens (protozoa, bacteria, viruses) in animals. The main group of insect vectors of both veterinary and human importance is Diptera including sand flies (*Phlebotomus* spp.), black flies (*Simulium* spp.), midges (*Culicoides* spp.), mosquitoes (belonging to several genera including *Aedes* spp., *Anopheles* spp., etc.), horseflies (*Tabanus* spp.), tsetse flies (*Glossina* spp.), and louse flies (*Hippobosca* spp.) [11].

Arthropod vectors are cold-blooded (ectothermic) and thus especially sensitive to climatic factors. Vectors require certain environmental characteristics that are unique for each type of organism. Mosquitoes, for instance, require humid conditions, whereas ticks can live in warm and dry climates [12].

Vector reproduction, survival, and distribution rely on environmental factors including temperature, humidity, and vegetation cover which are variable throughout the year and influence vector activity (biting rate). In the same country, a VBD prevalence in humans can vary from one season to another and from one area to another in the same period. In animals, VBD prevalence when transmitted by flies is linked to the intensity of animal-fly contact determined by the abundance and density of vectors, which is determined by climatic conditions. However, climate is only one of many factors influencing vector distribution and vector activity.

Several environmental components (vegetation, climate, geology) define the geographic area within which transmission takes place for a particular vector-host-pathogen system [13]. Impact of environmental factors on different pathogens and vectors is diverse and specific to individual vector-pathogen combinations.

Numerous viruses, bacteria, protozoa, and helminths have been found to require a hematophagous (bloodsucking) arthropod for transmission between vertebrate hosts. But, not all blood-sucking arthropods are vectors (transmitters) of disease agents [14]. Besides, vectors are not exclusive to any particular pathogen and can not only transmit more than one disease, but they can do so at the same time [11]. The ability and likelihood that a vector transmits a pathogen to a vertebrate host depends on a numerous of factors. These include the ability of an ingested pathogen to survive and multiply in the body of the vector and its ability to be transmitted during a subsequent blood meal to a vertebrate. Other determining factors are the number of pathogen ingested by the vector, the density of vectors in the environment, and its feeding preferences. The relationship between a specific vector and its preferred host is usually stable, but it can change, for various reasons including unavailability of the preferred host [15]. Another distinction that can be made is between primary and secondary vectors that is recognized for their importance in a disease transmission, keeping in mind that a known secondary vector on a global level can be seen as a primary vector on a local level.

An arthropod may transmit disease agents from one vertebrate host to another in two different ways, but the most important type of transmission is biological transmission. Biological transmission refers to morphological and physiological changes that a pathogen undergoes before its transmission from one host (vector) to another (vertebrate) belonging to different taxa. That modification enables pathogen's adaptation in the organism of its two hosts and occurs during a cycle called development cycle. The ability to transmit a pathogen biologically varies greatly among species of arthropods and even among geographical strains within a species [16]. Four types of biological transmission have been described according to the type of biological development the pathogen undergoes in the body of the arthropod vector:

i. Propagative transmission occurs when the organism ingested with the blood meal undergoes simple multiplication in the body of the arthropod.

ii. Cyclopropagative transmission in which the pathogen undergoes a developmental cycle (changes from one stage to another) as well as multiplication in the body of the arthropod.

iii. Cyclodevelopmental transmission: the pathogen undergoes developmental changes from one stage to another but does not multiply.

iv. Vertical and direct transmission occurs either via the transovarial route or by infection of eggs from female after oviposition, both leading to venereal transmission.

The second way of transmission called mechanical transmission consists of a simple transfer of the organism on contaminated mouthparts or other body parts. No multiplication or developmental change of the pathogen on or in the arthropod takes place during this type of transmission. The relative infection of the vector is usually of short duration in such cases because the vector is mere a pathogen carrier. Successful mechanical transmission depends on the degree of contact insects have with the vertebrate hosts and on feeding behavior [16].

The vector and pathogen interactions greatly affect the dynamics of VBDs and explain many of the particular characteristics of each infection and its epidemics. Vector-borne pathogen transmission occurs when host, vector, and pathogen interact in space and time within a permissive environment. The ability of arthropods to transmit a disease agent depends on many complex factors including ecological changes, either natural or human-induced, climate change, habitat destruction, and changes in population density/distribution. It has been assumed that observed changes in temperature, rainfall, and humidity that are occurring under different climate change scenarios affect the biology and ecology of vectors and intermediate hosts and consequently the risk of disease transmission [17].

Temperature can affect both distribution of vectors and effectiveness of pathogen transmission through vectors. Ticks exhibit strong seasonal dependence of mortality and disease transmission, which can be related to temperature and vegetation conditions [18]. Not all changes in climate favor vectors. Extreme temperatures can act positively or negatively on their development cycle. Many vectors need water for their maturation. Some species lay eggs directly onto the water surface, and others need moist substrates near water. For the first group, heavy rains and floods are directly related to high vector density, while for insects needing micro-environment next to backwater and ponds, floods are inadequate for their development. Such a negative effect is also observed in temperate zones during colder and longer winters [19, 20]. When temperature increases, larvae reach maturity within a very short time, and that enhances its capacity to produce more offspring. The vector biting rate increases and, consequently, intensity of disease transmission because the blood ingested by the hematophagous insect is rapidly digested [21].

In tropical areas, although climate patterns, particularly temperature and rainfall trends, have direct effects on VBD transmission [22], the actual magnitude and spatial extent of VBDs within regions are also governed by several non-climatic factors including epidemiological, environmental, social, economic and demographic factors [19, 23, 24]. Human interventions on the environment through urbanization, deforestation, domestic and industrial use of chemical products, migration, modern agricultural systems, and increase in the emissions of greenhouse gases are known to also influence dynamics of VBDs [11]. Global warming offers better conditions for the development of some vectors by decreasing temperatures and reducing diurnal and nocturnal ranges [19]. There are countries where environmental conditions are not so favorable for certain vector populations, but immigration allows them to persist [25].

3. Emerging vector-borne diseases and epidemics

Several infectious diseases have emerged during the last four decades in both animal and human health sectors. Other diseases considered to have been under control or quiescent have resurged, often spreading to geographical areas in which they had not been previously found, due to various reasons including movement of vertebrate host, human, or animal. Pathogens of some of them further need a "carrier" to switch from an infected host to another and are vector-borne.

Emerging diseases is usually defined as infections that have newly appeared in the population or have existed but are rapidly increasing in incidence or geographic range [26]. An author [27] defines the word "emerging diseases" as all entities comprising resurgent or recurrent known diseases (usually caused by "new" or mutated previously known agents), truly new diseases to man, but caused by animal pathogens, and syndromes caused by new agents easily detected through advances in research and development. More simply, a disease is recognized as "new" when its clinical signs are distinct from other known diseases [28].

Diseases transmitted by ticks and insects called vector-borne diseases (VBDs) are a growing threat all over the world. In Central Africa, VBDs have been playing a particularly important role because many of them are endemic and the burden of VBDs continues to be very heavy both in animal and human health sectors.

Over decades, trypanosomiasis has been described as a major animal disease constraint to livestock production in sub-Saharan Africa. It is estimated that animal trypanosomoses significantly reduce the number of cattle: 37% in subhumid zone to 70% in humid zone. The meat production is reduced by 5–30%, milk production from 10 to 40%. The cost of these losses is estimated at 1338 millions of US dollars. Production work for draft oxen is reduced by 33%. Although losses due to morbidity (loss of weight, decrease in reproductive performance, increase in interrelated diseases, etc.) are more difficult to evaluate, the total agricultural production would be reduced by 2–10% in the risk zone [29]. The disease is cyclically transmitted by tsetse flies, and its occurrence normally overlaps tsetse fly distribution across Africa: approximately 11 million km^2 between 15°N and 29°S. The incidence and severity of trypanosomiasis in livestock are closely related to the species of *Glossina* present in the area. Although species of the subgenus *Austenina* (fusca group) are mainly incriminated in the transmission of animal trypanosomiasis and in a lesser extent species of the subgenus *Glossina* (morsitans group), all 31 species of tsetse flies can transmit animal trypanosomiasis, depending to their individual vector competence and capacity. But species of animal trypanosomes and even trypanosome infectious rate is different from one locality to another.

Ecology of tsetse flies is highly correlated to environmental factors, including temperature (20–30°C) and humidity. In general, higher temperatures (>38°C) are not suitable for the survival of adults, and lower temperatures (>17°C) do not allow immature stages to complete their development cycle. Therefore, tsetse flies are absent in hot and dry areas. But tsetse flies are not the only vectors of animal trypanosomiasis. Tabanids and stomoxes can also transmit that disease, though mechanically [30], and contribute to its spread even in tsetse-free areas.

In Central Africa countries, mainly in Cameroon and Chad, African animal trypanosomiasis (AAT), also called Nagana, has been so far found in savannah areas, and 70 percent of drug farmers' expenses are for trypanocides. There are important seasonal variations in the degree of risk to which livestock are exposed in areas that have pronounced variation between wet and dry seasons. During pronounced dry seasons, there is a general regression in the distribution of tsetse particularly if the dry season is also hot. The burning, usual in the savannah lands, accelerates the diminution in the extent of suitable tsetse habitats, and these have been seen as natural tsetse control measures. Farmers with trypanosomiasis-sensitive cattle take advantage of these and implement a rotational breeding system based on seasonal variations.

Recently, there has been a gradual expansion of AAT in known tsetse-free areas in Cameroon and Chad. Indeed, the Ministry in charge of livestock of Cameroon, through it national office against Glossina (MSEG) has noticed an increase incidence of AAT due to *T. brucei* and *T. vivax* in a tsetse-free area: the locality of Pette (10.97 N, 14.50S), Diamare Division, Far North region. Disease is mechanically transmitted and maintained by tabanids, present in the area (MSEG activity report, unpublished). The emergence of AAT in Pette (Cameroon) and in some areas beyond the tsetse belt of South-East of Chad is not the only emerging VBD that face countries of Central Africa.

In April 2019, Cameroon veterinarian services notified sudden morbidity of donkeys and mortalities of horses in two divisions of the Far North region detected by the national network surveillance of animal diseases (RESCAM report, unpublished). After laboratory investigation, African horse sickness (AHS) was confirmed. It is the first time that disease is notified in that country. In the meantime, AHS was also officially declared in Chad, while it was 3 months ago reported by veterinarian services of Nigeria. AHS is a vector-borne viral disease transmitted by *Culicoides* spp. Following the first notification of that disease in Cameroon, three divisions of the same region and two of the adjacent North region were also affected. It was an epidemic.

AHS occurs across a wide range of biotic and abiotic parameters that relate to interactions among host, pathogen, vector, and environment. *Culicoides* spp. remains the least studied of the major dipteran vector groups despite their veterinary and economic importance. The small size and fragility of *Culicoides* spp. and their limited direct impact on public health occurring through nuisance biting inflicted by female adults [31] could explain that lack of attention. Nevertheless, more than 1400 species have been described worldwide (except Antarctica and New Zealand) in the genus *Culicoides*. Geographical distribution and ecology of these holometabolous flies rely on the existence of moisture-rich habitats that are necessary for their development cycle [32]. Thus, the occurrence of AHS is preceded by seasons of heavy rain that alternate with hot and dry climatic conditions, which favor transmission by the insect vector [33].

Arboviruses circulate among wild animals, and many can be transmitted to humans and agriculturally important domestic animals through a process known as spillover [34]. The emergence of African horse sickness in Central Africa could be due the movement of reservoir animals from one area to another or the importation of vectors which can bring with it new diseases with great adaptability in the part of the pathogen and the vector.

Epidemics of vector-borne disease have arisen from specific conditions occurring within the context of the large-scale drivers of infectious disease emergence. Global climate change has been assumed to lead to an increase of vector-borne infectious diseases and disease outbreaks. It could affect the range and population of pathogens, host, and vectors and transmission season [21]. Changes in ecosystem lead to the increase of population in natural hosts or vectors for certain emerging infectious disease. These factors are becoming increasingly prevalent [35], suggesting that infections will continue to emerge and probably increase. In fact, although not endemic, cases of Rift Valley fever (RVF) are increasingly detected by the national veterinary laboratory of Cameroon in ruminants mainly in the North region. Evidence of the circulation of RVF virus in Cameroon was demonstrated in 2017 on samples collected in years 2013 and 2014 on small ruminants and cattle [36]. Few years ago, some authors had shown the presence of RVF virus in goats in forests of South region Cameroon [37] and others in domestic ruminants in Chad [38]. According to them, the presence of RVF virus antibodies in domestic animals suggests that this virus may also be circulating in human populations, despite the absence of reports.

In addition to being a vector-borne disease, RVF can be transmitted to humans (by nonvector means) and causes hemorrhagic fever, encephalitis, and mortalities. Mosquitoes are main "transmitters" between animals, while humans can contract disease after a direct contact with infected bodily fluids or tissues of infected animals. Human infection after mosquito bites is mild or asymptomatic [39, 40]. Epidemiological patterns of that disease differ from one area to another. In East Africa, several RVF outbreaks have been linked to prolonged heavy rainfall [41], whereas in West Africa, outbreaks usually occur during years of normal or poor rainfall [42, 43]. Such studies have not yet been conducted in Cameroon.

A change in the geographical distribution of a vector-borne disease is related to a change in the distribution of its vector and/or the pathogen through human or animal movement. Due to climate change leading to unsuitable environmental conditions for its survival, vector moves and thus shifts its geographical range of distribution. But the mere presence of a vector does not necessarily mean presence of a disease. Human African trypanosomiasis due to *Trypanosoma brucei gambiense* has been described to not exist in some areas where the main vector *Glossina palpalis* prevails. Whatever it is, factors contributing to emerging and re-emerging zoonoses can be divided into two groups: intrinsic factors and extrinsic factors. Intrinsic factors are factors that lead to the emergence of new pathogens; and extrinsic factors are those related to environmental changes or human behavior including deforestation, urbanization, and agropastoral activities.

Deforestation provides new ecological niches and conditions for proliferation of newly arriving and/or adaptive vectors and their pathogens. Agriculture and livestock production create a favorable habitat for pathogens and their respective host vectors. Humans are exposed to new pathogens during their daily activities and animals, mainly in pastures. Because of the scarcity of their preferential vertebrate host, some vectors display a conversion from a primarily zoophilic (bites to animal) to primarily anthropophilic (bites to human) orientation [44]. That hypothesis has been argued to explain prevalence of human African trypanosomiasis due to *Trypanosoma brucei rhodesiense* and transmitted by *Glossina morsitans,* assumed to be linked to the disappearance of wild mammals. In the same way, water control projects create new breeding habitats for insects, their larvae, and their pathogens. The construction of new roads provides access for new human, livestock, vector, and pathogen populations [45].

In addition to increased public health response, a better understanding of the epidemiology of VBDs is needed to identify the drivers of these epidemics and inform the public health response.

4. Impact of emerging vector-borne diseases in Central Africa

The relative significance of most VBDs in livestock is extremely difficult to quantify, because there is no study about their socioeconomic impact. However, it is known certain VBDs do have a particularly significant impact on food production.

It has been assumed that animal trypanosomiasis that prevents the use of draft animals has serious impact on land use. Cattle farming is practiced mainly in dry areas where tsetse flies cannot survive and also where there are not large areas of pasture. The consequence of that is overstocking and land degradation [46]. Emerging of trypanosomiasis in dry area maintained by mechanical transmission further impoverishes local populations. In the locality of Pette (Cameroon) where such a situation has occurred, farmers have adopted a night pasture system. That means animals are kept from tabanids bites in day time inside cowsheds and brought to grazing areas overnight when tabanids are inactive. The consequence is that not only cattle are undernourished but also local populations who live from their farm production. With the non-governmental organizations' (NGO) support aiming at increasing their productivity and income, farmers have implemented sustainable agricultural methods to preserve natural resources. These methods involved water management in irrigation scheme. Irrigation is needed not only to grow food for human consumption but also for livestock feed and provides suitable environment conditions for mosquitoes and midges. So there was no surprise about the occurrence of others VBD even among populations. For example, storage dams built in irrigation schemes to improve food security in the

Far North region of Cameroon have been shown to provide suitable habitats for the aquatic snails, the intermediate hosts of schistosome parasites. As these areas are also used by local populations for other domestic purposes including laundry and swimming by children, they are exposed to the emerging schistosomiasis which becomes endemic. Vector-borne diseases are a threat to animal and human health, animal welfare, and trade. Impact of RVF on domestic markets and international trade is probably greater than the direct mortality and production impacts of the disease. Trade restriction still remains one of RVF's major economic effects that prevent farmers to declare to veterinary services first cases of RVF or report an outbreak when occurring. This can delay implementation of control measures until the disease spreads to neighboring areas, sometimes across borders, and creates an international epidemic [47].

Morbidity and mortality from AHS in the Far North and North regions of Cameroon have constrained the draft power of working equine, mainly donkeys, thereby affecting food security, poverty alleviation, and gender equality. AHS virus is also a major threat to equine sport. Several races took place across Nigeria, Cameroon, and Chad and within countries, across regions. Economic value of horse racing, though not evaluated, is considerable. The value of horses as companion animals is less well defined but can provide physical and psychological benefits to owners and riders. Epidemics of AHS were controlled by quarantine of equines moving from endemic and epidemic AHS regions or country to virus-free areas, vaccination, and stabling. Because of that equine movement restriction, local farmers underwent serious losses in their income.

In Senegal, a study conducted on the economic impact of AHS after the outbreak of 2007 revealed a loss of US$ 1,793,581.596 following a morbidity rate of 0.26% and a mortality rate of 0.23% of equine [48]. Following the 2006–2007 RVF outbreaks in Kenya, a socioeconomic study has shown significant financial losses along the livestock value chain. The value of these losses ranged from US$ 1500–8900 by actors, depending on the nature of product lost. For livestock producers, that loss of milk production was caused by abortion. Livestock traders lost sales because of animal deaths, while slaughterhouses closed, and butchers stopped their activities because of fewer animals being killed and marketed [49].

The public health impact of VBDs is induced by pathogens that can be transmitted to humans. Although major livestock outbreaks precede human epidemics, most of the pronounced RVF events in history were first diagnosed in humans. The amplifying livestock epizootics were recognized only after the disease was noticed in the human population [47]. Such a situation is not yet documented in Central Africa. Nevertheless, inter-epidemic serological evidence in human was demonstrated in Gabon [50] which shared with Cameroon the second large forest of the world: the Congo Basin. That forest harbors more than 10 species of forest ungulates, and RVF virus has been detected from a number of wild African ungulates [51]. The potentiality of RVF virus and other pathogens to migrate between ungulates, domestic animals, and humans could have considerable effects not only on animal production but also on human health.

The impact of vector-borne diseases on human health, animal health, and trade as well as the transboundary nature of these diseases means there is a need for regional coordination and cooperation to address challenges.

5. Suggesting approach to control emerging vector-borne diseases

The complex epidemiology of vector-borne diseases creates significant challenges in the design and delivery of prevention and control strategies, especially in sight of rapid social and environmental changes.

VBD management must be based on realistic and achievable objectives such as reducing the burden or interrupting transmission cycle of the disease. The most successful approaches to the management of VBDs of importance to livestock and humans are often multifaceted and include awareness creation, treatment, or vaccines as well as efforts to reduce the population of the vector. But for a number of livestock vector-borne pathogens, no vaccines are available.

It has been suggested to use vaccination as a control strategy to limit the circulation of RFV virus in enzootic areas and to prevent epidemics in free areas. Vaccination is most effective when used in conjunction with other control strategies like movement restriction and sanitary slaughter [52]. In contrary, immunological control of African trypanosomiasis is not possible yet because of the antigenic variation of trypanosomiasis agents. Therefore, trypanocides are widely used to control AAT in cattle. However, no new veterinary drugs for the treatment of AAT have been released since 1985 [53], and there is increasing resistance to the existing trypanocides. Mitigation risk of exposure goes through avoiding grazing in infested areas and use of prophylactic trypanocides. Trypanotolerant breeds and crossbreds are also recommended [54] but not appreciated by farmers because of their low production rate.

Vector control can also be part of an overall strategy for reducing host-pathogen contact through the vector. This strategy has provided great success in Burkina Faso [55, 56] and Cameroon [57, 58] against AAT. Larviciding measures at mosquito breeding sites are the most effective form of vector control against RVF if breeding sites can be clearly identified and are limited in size and extent. Keep in mind that during periods of flooding, the number and extent of breeding sites is usually too high for larviciding measures to be feasible [59].

Training and education is paramount. Communities need to be educated both on the impact of zoonotic diseases and control methods of VBDs. It has been assumed that education of inhabitants on the pathological impacts of AAT on animal health and peasant economy will ease their cooperation for control activities that will guarantee and ensure sustainability and success of control measures [60].

Veterinarian capacities should be built on one health approach that helps to address the complexities of VBDs and its associated impacts. The One Health concept signifies a collaborative, multidisciplinary, and holistic approach, looking at optimizing animal, environmental, and human health, which are interdependent on each other [61]. Bringing animal health, human health, and environmental actors and partners together within the type of One Health program optimizes the use of scarce resources and could achieve cost-effective benefits for all components targeting conservation and human well-being. Epidemics control and more effective mitigation of impacts can be achieved by coordinated actions involving human, animal, and environmental health to prevent, detect, and respond to animal and human diseases as well as infected vector populations. Environmental risk assessment and early detection of pathogen in livestock is the best approach to protect human health.

But this cannot be successful without an effective surveillance system. Risks related to the transmission of disease need to be determined including evaluation of the potential spread to new areas or the introduction of exotic species or diseases. Predicting outbreaks and early detection are useful tools to mitigate animal health and economic impacts. Surveillance and preparedness should be implemented in a multi-sectoral approach that fully integrates animal and human health sectors, epidemiology, wildlife, and environment. A holistic approach that transcends disciplines such as joint-risk assessment, joint investigation, and response is essential to implement risk-based surveillance and build overall response capacity.

Countries should put in place-specific measures when dealing with transboundary animal diseases and targeting cross-border ecosystems. Emerging vector-borne disease means that the lack of zoo sanitary precautions at national borders including sustainable vector control has contributed to its spread from a neighbor country, especially in light of predominant climate change scenarios.

6. Conclusions

There is a regional emergence or re-emergence and expanding geographical distribution of vector-borne diseases in Central Africa, with an increased frequency of epidemic transmission. Central Africa is a hotspot of emergence or re-emergence disease. Impact of these diseases goes beyond animal mortality and mortality that seriously affects animal production and prevents poverty reduction to reach and threat human health. Reversing the trend of emergent/resurgent vector-borne diseases is very challenging. One Health-oriented collaborations among professionals working in diverse sectors such as animal health, human health, public health, entomology, and animal production will contribute to overcome the challenges faced by the sustainability of control of VBDs.

Conflict of interest

The authors declare no conflict of interest.

Author details

Lisette Kohagne Tongue[1*] and Arouna Njayou Ngapagna[2]

1 Fight Against Parasitoses Association (APLP), Yaounde, Cameroon

2 Higher Institute of Health Sciences of Bangangté (ISSS), Bangangté

*Address all correspondence to: lisetteappmv@yahoo.fr

IntechOpen

© 2019 The Author(s). Licensee IntechOpen. This chapter is distributed under the terms of the Creative Commons Attribution License (http://creativecommons.org/licenses/by/3.0), which permits unrestricted use, distribution, and reproduction in any medium, provided the original work is properly cited.

References

[1] Winrock International. Assessment of Animal Agriculture in Sub-Saharan Africa. Morrilton, Arkansas, USA: Winrock International; 1992. 125p

[2] Otte MJ, Chilonda P. Cattle and Small Ruminant Production Systems in Sub-Saharan Africa: A Systematic Review. Rome: Food and Agriculture Organization of the United Nations; 2018. p. 2002. 105p

[3] Robinson TP, Thornton PK, Franceschini G, Kruska RL, Chiozza F, Notenbaert A, et al. Global Livestock Production Systems. Rome: Food and Agriculture Organization of the United Nations (FAO) and International Livestock Research Institute (ILRI); 2011. 152p

[4] Jahnke HE. Livestock Production Systems and Livestock Development in Tropical Africa. Kiel: Wissenschaftsverlag Vauk; 1982

[5] ILCA. Livestock Production in the Subhumid Zone of Western Africa. Systems Study 2. Addis Ababa: ILCA; 1979

[6] Bellan MF. Ecofloristic Zones and Global Ecological Zoning of Africa, South America and Tropical Asia. Rome: Food and Agriculture Organization of the United Nations; 2000

[7] Wilson RT. Livestock Production Systems. London: Macmillan; 1995. 141p

[8] Toledo A, Olmeda AS, Escudero R, Jado I, Valcarcel F, Casado-Nistal MA, et al. Tick-borne zoonotic bacteria in ticks collected from Central Spain. The American Journal of Tropical Medicine and Hygiene. 2009;**81**(1):67-74

[9] Liyanaarachchi DR, Rajakaruna RS, Dikkumbura AW, Rajapakse RP. Ticks infesting wild and domestic animals and humans of Sri Lanka with new host records. Acta Tropica. 2015;**142**(2):64-70

[10] Ostfeld RS, Price A, Hornbostel VL, Benjamin MA, Keesing M. Controlling ticks and tick-borne zoonoses with biological and chemical agents. Bioscience. 2006;**56**(5):383-394

[11] Leitner WW, Wali T, Kincaid R, Costero-Saint Denis A. Arthropod vectors and disease transmission: Translational aspects. PLoS Neglected Tropical Diseases. 2015;**9**(11):e0004107. DOI: 10.1371/journal.pntd.0004107

[12] Demma LJ, Traeger MS, Nicholson WL, Paddock CD, Blau DM, et al. Rocky mountain spotted fever from an unexpected tick vector in Arizona. The New England Journal of Medicine. 2005;**353**(6):587-603

[13] Reisen WK. Landscape epidemiology of vector-borne diseases. Annual Review of Entomology. 2010;**55**:461-483. DOI: 10.1146/annurev-ento-112408-085419

[14] EurNEgVEc One Health Dictionary. A Product of a COST Action TD1303 European Network for Neglected Vectors and Vector Borne Infections. Available from: https://www.eurnegvec.org/publications/other/EurNegVecDictionary.pdf

[15] Verwoerd DW. Definition of a vector and a vector-borne disease. Revue Scientifique et Technique. 2015;**34**(1):29-31

[16] Gubler DJ. Vector-borne diseases. Revue Scientifique et Technique. 2009;**28**(2):583-588

[17] Mohms Fiji. In: Proceedings of the Workshop on Climate Change and Vector-Borne Disease. Suva, Fiji; 10-12 February 2015. 23p

[18] Randolph SE. Ticks and tick-borne disease systems in space and from space. Advances in Parasitology. 2000;**47**:217-240

[19] Reiter P. Climate change and mosquito-borne diseases. Environmental Health Perspectives. 2001;**109**(1):141-160

[20] Kovats RS, Campbell-Lendrum DH, McMichael AJ, Woodward A, Cox JS, et al. Early effects on climate change: Do they include changes in vector-borne diseases? Royal Society of London B. 2001;**356**:1057-1068

[21] Githeko AK, Lindsay SW, Confalonieri UE, et al. Climate change and vector-borne diseases: A regional analysis. Bulletin of the World Health Organization. 2000;**78**:1136-1147

[22] Lafferty KD. The ecology of climate change and infectious diseases. Ecology. 2009;**9**:888-900. DOI: 10.1890/08-0079.1

[23] Qi Q, Guerra CA, Moyes CL, Elyazar IR, Gething PW, Hay SI, et al. The effects of urbanization on global *Plasmodium vivax* malaria transmission. Malaria Journal. 2012;**11**:403. DOI: 10.1186/1475-2875-11-403

[24] Yang HM, Ferreira MU. Assessing the effects of global warming and local social and economic conditions on the malaria transmission. Revista de Saúde Pública. 2000;**34**:214-222. DOI: 10.1590/S0034-89102000000300002

[25] Rascolau G, Pontier D, Menu F, Gourbiere S. Emergence and prevalence of human vector-borne diseases in sink vector populations. PLoS One. 2012;7(5):e36858. DOI: 10.1371/journal.pone.0036858

[26] Morse SS. Factors in the emergence of infectious diseases. Emerging Infectious Diseases. 1995;**1**(1):7-15

[27] Kilbourne ED. The emergence of "emerging diseases": A lesson in holistic epidemiology. Mount Sinai Journal of Medicine. 1996;**63**(3-4):159-166

[28] Levins R, Awerbuch T, Brinkman U, Eckardtt I, Epstein P, et al. The emergence of new diseases. American Scientist. 1994;**82**:52-60

[29] Swallow BM. Impacts of Trypanosomosis in African Agriculture. Rome: Food and Agriculture Organization of the United Nations, PAAT Technical and Scientific series; 2000

[30] Desquesnes M, Dia ML. Trypanosoma vivax: Mechanical transmission in cattle by one of the most common African tabanids, *Atylotus agrestis*. Experimental Parasitology. 2003;**103**(1-2):35-43. DOI: 10.1016/S0014-4894(03)00067-5

[31] Carpenter S, Martin H, Groschup C, Garros M, Felippe-Bauer L, Purse BV. Culicoides biting midges, arboviruses and public health in Europe. Antiviral Research. 2013;**2013**(100):102-113

[32] Mellor P, Boorman J, Baylis M. Culicoides biting midges: Their role as arbovirus vectors. Annual Review of Entomology. 2000;**45**:307-340

[33] Carpenter S, Philip S, Mellor G, Fall A, Garros C, Venter GJ. African horse sickness virus: History, transmission, and current status. Annual Review of Entomology. 2017;**62**:343-358

[34] Huang Y-JS, Higgs S, Vanlandingham DL. Arbovirus-mosquito vector-host interactions and the impact on transmission and disease pathogenesis of arboviruses. Frontiers in Microbiology. 2019;**10**:22. DOI: 10.3389/fmicb.2019.00022.36

[35] Miled Pherez F. Factors affecting the emergence and prevalence of vector borne infections (VBI) and the role of vertical transmission (VT). Journal of Vector Borne Diseases. 2007;**44**:157-163

[36] Rissmann M, Eiden M, Wade A, Poueme R, Abdoulkadiri S, Unger H, et al. Evidence for enzootic circulation of Rift Valley fever virus among livestock in Cameroon. Acta Tropica. 2017;**17**:7-13

[37] LeBreton M, Umlauf S, Djoko CF, Daszak P, Burke DS, Yemgai Kwenkam P, et al. Rift Valley fever in goats, Cameroon. Emerging Infectious Diseases. 2006;**12**(4):702-703. DOI: 10.3201/eid1204.051428

[38] Ringot D, Durand JP, Tolou H, Boutin JP, Davoust B. Rift Valley fever in Chad. Emerging Infectious Diseases. 2004;**10**(5):945-947

[39] Hoogstraal H, Meegan JM, Khalil GM, Adham FK. The Rift Valley fever epizootic in Egypt 1977-1978. Ecological and entomological studies. Transactions of the Royal Society of Tropical Medicine and Hygiene. 1979;**73**(6):624-629

[40] Pepin M, Bouloy M, Bird BH, Kemp A, Paweska J. Rift Valley fever virus (Bunyaviridae: Phlebovirus): An update on pathogenesis, molecular epidemiology, vectors, diagnostics and prevention. Veterinary Research. 2010;**41**(6):61. DOI: 10.1051/vetres/2010033

[41] Sang R, Arum S, Chepkorir E, Mosomtai G, Tigoi C, Sigei F, et al. Distribution and abundance of key vectors of Rift Valley fever and other arboviruses in two ecologically distinct counties in Kenya. PLoS Neglected Tropical Diseases. 2017;**11**(2):e0005341. DOI: 10.1371/journal.pntd.0005341

[42] Fontenille D, Traore-Lamizana M, Diallo M, Thonnon J, Digoutte JP, Zeller HG. New vectors of Rift Valley fever in West Africa. Emerging Infectious Diseases. 1998;**4**(2):289-293

[43] Chevalier V, Lancelot R, Thiongane Y, Sall B, Diaité A, Mondet B. Rift Valley fever in small ruminants, Senegal, 2003. Emerging Infectious Diseases. 2005;**11**(11):1693-1700

[44] Ozer N. Emerging vector-borne diseases in a changing environment. Turkish Journal of Biology. 2005;**29**:125-135

[45] Patz JA, Graczyk TK, Geller N, et al. Effects of environmental change on emerging parasitic diseases. International Journal for Parasitology. 2000;**30**(1-11):1946-1955

[46] Service MW. Agricultural development and arthropod-borne diseases: A review. Revista de Saúde Pública. 1991;**25**(3):165-178

[47] Mariner J. Rift Valley Fever Surveillance. Rome: FAO Animal Production and Health Manual Food and Agriculture Organization of the United Nations (FAO); 2018. 80p

[48] Wombou Toukam CM, Ly C, Akakpo AJ. Economic impact of African horse sickness in Senegal: The outbreak of 2007. ISVEE/851; 2007. 5p

[49] Rich KM, Wanyoike F. An assessment of the regional and national socio-economic impacts of the 2007 Rift Valley fever outbreak in Kenya. The American Journal of Tropical Medicine and Hygiene. 2010;**83**(2 Suppl):52-57

[50] Pourrut X, Nkoghé D, Souris M, Paupy C, Paweska J, Padilla C, et al. Rift Valley fever virus seroprevalence in human rural populations of Gabon. PLoS Neglected Tropical Diseases. 2010;**4**(7):e763

[51] Paweska JT, Smith SJ, Wright IM, Williams R, Cohen AS, Van Dijk AA.

Indirect enzyme-linked immunosorbent assay for the detection of antibody against Rift Valley fever virus in domestic and wild ruminant sera. The Onderstepoort Journal of Veterinary Research. 2003;**70**:49-64

[52] FAO. Rift Valley Fever: Vigilance Needed in the Coming Months. Rome: Empres Watch; 2012

[53] Anene BM, Onah DN, Nawa Y. Drug resistance in pathogenic African trypanosomes: What hopes for the future? Veterinary Parasitology. 2001;**96**(2):83-100

[54] Meyer A, Holt HR, Selby R, Guitian J. Past and ongoing tsetse and animal trypanosomiasis control operations in five African countries: A systematic review. PLoS Neglected Tropical Diseases. 2016;**10**(12):e0005247. DOI: 10.1371/journal.pntd.0005247

[55] Merot P, Politzar H, Tamboura I, Cuisance D. Results of a control campaign against river tsetse flies in Burkina using deltamethrin impregnated screens. Revue d'Élevage et de Médecine Vétérinaire des Pays Tropicaux. 1984;**37**(2):175-184

[56] Holt H, Selby R, Mumba C, Napier G, Guitian J. Assessment of animal African trypanosomiasis (AAT) vulnerability in cattle-owning communities of sub-Saharan Africa. Parasites & Vectors. 2016;**9**(1):1

[57] Mamoudou A, Zoli A, Van den Bossche P, Delespaux V, Cuisance D, Geerts S. Half a century of tsetse and animal trypanosomosis control on the Adamawa plateau in Cameroon. Revue d'Élevage et de Médecine Vétérinaire des Pays Tropicaux. 2009;**62**(1):33-38

[58] Tanenbe C, Gambo H, Musongong G, Boris O, Achukwi M. Prévalence de la trypanosomose bovine dans les départements du Faro et Déo, et de la Vina au Cameroun: bilan de vingt années de lutte contre les glossines. Revue d'Élevage et de Médecine Vétérinaire des Pays Tropicaux. 2010;**63**(3-4):63-69

[59] WHO. 2017. Rift valley fever. Available from: http://www.who.int/mediacentre/factsheets/fs207/en/ [Retrieved: 20 February 2017]

[60] Simo G, Rayaisse JB. Challenges facing the elimination of sleeping sickness in west and Central Africa: Sustainable control of animal trypanosomiasis as an indispensable approach to achieve the goal. Parasites & Vectors. 2015;**8**:640. DOI: 10.1186/s13071-015-1254-y

[61] McConnell I. One health in the context of medical and veterinary education. Revue Scientifique et Technique. 2014;**33**(2):651-657

Chapter 3

Mosquito-Borne Viral Diseases: Control and Prevention in the Genomics Era

Vagner Fonseca, Joilson Xavier, San Emmanuel James, Tulio de Oliveira, Ana Maria Bispo de Filippis, Luiz Carlos Junior Alcantara and Marta Giovanetti

Abstract

Mosquito-borne viral diseases are infections transmitted by the bite of infected mosquitoes. The burden of these diseases is highest in tropical and subtropical areas and they disproportionately affect the poorest populations. Since 2014, major outbreaks of dengue, chikungunya, yellow fever and Zika have afflicted populations and overwhelmed health systems in many countries. Distribution of mosquito-borne diseases is determined by complex demographic, environmental and social factors, causing diseases to emerge in countries where they were previously unknown. Coupling genomic diagnostics and epidemiology to innovative digital disease detection platforms raises the possibility of an open, global, digital pathogen surveillance system. Considering pathogen surveillance in mind, real-time sequencing, bioinformatics tools and the combination of genomic and epidemiological data from viral infections can give essential information for understanding the past and the future of an epidemic, making possible to establish an effective surveillance framework on tracking the spread of infections to other geographic regions.

Keywords: mosquito-borne viral diseases, arboviral infections, genomics epidemiology, next-generation sequencing, genomic surveillance, viral pathogens

1. Introduction

Mosquito-borne viral diseases have lately integrated worldwide headlines since the emergence of arbovirus outbreaks in big urban areas. According to the World Health Organization, more than 17% of all infectious diseases registered worldwide are represented by vector-borne diseases, and they account for more than 700,000 deaths annually [1]. Due to this scenario of increasing cases number and expansion to new areas, the spread of infectious diseases was listed second in the top 10 risks in term of impact according to the Global Risks 2015 report [2].

Mosquitos of the genus *Aedes* have been responsible for the emergence and re-emergence of many arboviral diseases worldwide [3]. The species *Aedes aegypti* is the main vector species responsible for the major arbovirus epidemics recorded in recent years [4]. The species *A. aegypti* and *A. albopictus* are possibly suitable to survive and establish in 215 countries/territories, and their expanding range is underlined by the increasing number of countries reporting transmission of mosquito-borne viruses. Transmissions of arboviruses, such as Zika, dengue, chikungunya, yellow fever, and Rift Valley fever, have been reported in 85, 111, 106, 43, and 39 countries, respectively [5]. Projections indicated that 3.83 billion people are living in areas prone to transmission of dengue and it is predicted that by 2050 large increases in dengue suitability will be seen in southern Africa and in the Sahel in West Africa [14]. Bhatt et al. projected the global burden of dengue around the world whose estimate indicated that 96 million dengue infections occur per year worldwide and this number represents infections that manifest at any level of the disease severity [6]. the Americas, comprising North and South America, registered more than 2 million dengue cases in 2016, and more than 1.4 million cases in 2019 [7]. For chikungunya fever, the Americas registered more than 94,000 cases in 2018, and in that same region, Zika fever accounted for more than 650,000 cases in 2016 [8, 9]. High number of cases of arboviral diseases was also registered in other regions in recent years, such as in the western pacific region where more than 375,000 suspected dengue cases were reported in 2016 [10]. In Africa, the government of Congo reported 6149 suspected cases of chikungunya until April 2019, and more than 13,000 chikungunya cases were reported in Sudan until October 2018 [11, 12]. The increasing in frequency and distribution of arboviral diseases in recent years represents a worrying burden not only for the public health system, but also for the economic sector [3]. Some estimates of the economic costs of arboviral infections have been made and for the case of dengue infections, it has been estimated that the median cost of all reported dengue hospital admissions registered in a municipality from Brazil was US$ 259.9 per hospitalization [13, 17]. Also, in Maldives, in the Indian Ocean, dengue fever represented a total cost of $3 million in 2015 [14]. Another estimate indicated that West Nile fever hospitalized cases in US represented a total cumulative cost of $778 million between 1999 and 2012 [15].

Dengue and chikungunya are two arboviral diseases present in the list of neglected tropical diseases from the World Health Organization. Neglected tropical diseases are a group of diseases that have received insufficient public attention, strive in tropical and subtropical areas, and strongly affect populations living in poverty [12]. It is argued that arboviruses can be considered a group of neglected tropical diseases, since they can have a long-lasting impact in the health and economic life of affected populations [16]. Some studies have argued that socioeconomic factors and land-use changes associated with the effects of climate change and global travel, and trade modulate the dynamics of expansion of emerging e re-emerging mosquito-borne diseases [17–20]. Movement of people between neighboring countries has been considered a good predictor for chikungunya spread in the Caribbean and Indian Ocean [14]. The expansion of the geographic distribution of arbovirus has significant negative impact on public health in many regions of the world. As measures to reduce such impacts, it has been argued about the relevance to public health of the implementation of a surveillance system that monitors virus diffusion and the appearance of new genetic variants [21]. In this sense, the use of genomic sequencing data and bioinformatics has been employed in the study of virus evolution, aiming to elucidate phylogenetic relationships and patterns of virus spread during an epidemic [22].

2. Genomic surveillance

Infectious diseases continue to be one of the leading causes of death worldwide [23] and pathogens such as viruses can evolve and spread rapidly, leading to the emergence of newly-mutated human pathogens, more virulent strains, as well as antibiotic and drug resistant organisms [24, 25]. In this context, genomic surveillance aims are to: (i) to perform global surveillance of pathogens using whole genome sequencing and (ii) to understand drug resistance, emergence and spread of viral pathogens. Several approaches have been developed and are widely used for the quick detection and identification of viral pathogens (i.e., diagnostics). Some of them are based on different serological and molecular strategies including, for example, assays based on real-time polymerase chain reaction [26]. Even though these kinds of approaches present high sensitivity and specificity for their purpose, they are more suitable for diagnostics only and cannot provide detailed genomic information [27].

Bearing these limitations in mind, the main point of developing new genomic surveillance tools is to answer the following inquiry: what sort of questions is important for genomic surveillance that cannot be addressed by conventional RT-qPCR or serology? (i) RT-qPCR assays do not allow genotype classification, neither does it help identify particular and/or characteristic transmission routes; (ii) RT-qPCR assays also do not allow to determine how fast a viral pathogen is being transmitted and in what direction it is spreading; (iii) serological and molecular assays also cannot help identify epidemiologically linked individuals, neither predict future outbreaks; and (iv) finally, serological and some molecular approaches cannot help to identify novel pathogenic agents and are, therefore, unsuitable for pathogen discovery [27].

Next generation sequencing (NGS) technologies produce significantly more raw data than other molecular diagnostic assays, including Sanger sequencing, and are also capable of informing not just pathogen diagnostics but also epidemiology [28]. This is why whole genome sequencing of viral genomes by using new technologies plays an important role in the fight against emerging and re-emerging epidemics [29, 30]. The availability of high-throughput sequencing has also provided immense insights into the ecology of health care-associated pathogens [31]. Therefore, real-time sequencing of entire pathogen genomes has become a standard and indispensable research tool for the critical role of genomic surveillance in the prevention and control of emerging infectious diseases [32], which justifies why NGS can be considered a powerful strategy that also allows the discovery of novel potential viral pathogens [33, 34].

Considering pathogen surveillance in mind, bioinformatics tools and the combination of genomic and epidemiological data from viral infections can give essential information for understanding the past and the future of an epidemic, because genomic data generated by real-time sequencing can provide important information on how and when viruses were introduced in a particular site, their pattern and determinants of dissemination in neighboring locations and the extent of genetic diversity, i.e., its dynamics, making it possible to establish an effective surveillance framework on tracking the spread of infections to other geographic regions [21, 22, 34]. In this context, recently established international networks for real-time, portable genomic sequencing, genomic surveillance and data analysis made it possible to monitor the evolution of viral genomes, to understand the origins of outbreaks and epidemics, to predict future outbreaks and to assist in the maintenance of updated diagnostic methods [33–35]. Additionally, genomic surveillance framework allows to determine, through genome sequencing, the real-time molecular epidemiology of viruses circulating and co-circulating in different regions in a specific area, and also to detect and characterize the early emergence of new pathogens in large urban centers, generating data that can inform outbreak control responses [27, 34]. Generated data regarding the molecular, epidemiological, phylogenetic and geographical aspects of circulating viral pathogens in a specific setting contribute to a better understanding of those viral infections in a national and international context, assuming an important role in solving issues relevance to Public Health [35]. As a result, studies involving more in-depth molecular and dispersion analysis of circulating pathogens may help the World Health Organization appropriately adopt measures to control epidemics and to monitor the dynamics and spreading of new viral strains. However, even though NGS has advantages over diagnostics routine, all of the different strategies and technologies, developed by Illumina, Thermo Scientific, Oxford Nanopore and others, are not yet considered a panacea. Remaining challenges include dealing with high data throughput, which requires sophisticated computational processing as well as the annotation of large amounts of sequencing data, high DNA or RNA input sample requirements (in some cases hundreds of nanograms), which often raises the need for previous PCR-based amplification approaches. On top of all this, there are relatively few researchers in the area with sufficient bioinformatics expertise and who are able to engage in near-patient or disease surveillance activities [35].

3. Bioinformatics tools and phylogenetic tools

The advent of next generation sequence (NGS) and advancements in bioinformatics present an opportunity to tap into new insights that are crucial to the establishment of an open, global digital surveillance system. NGS technologies have enabled the production and deposit of vast amounts of whole genomes into public repositories [36–38] ushering the field of genomics into era of big data. This has in turn increased the scale of genomic studies from the analysis of single or few genomes to an ever-increasing large number of genomes [39, 40].

Toward the development of global surveillance system, bioinformatics provides the tools to answer pertinent questions including the identification of organisms responsible for an outbreak, the source of an outbreak and evolutionary information of pathogens crucial for understanding the unique phenotypes such as drug resistance, virulence and disease outcome.

Several bioinformatics tools and pipelines have been developed to facilitate the processing, analysis and visualization of these data in order to derive useful information from it [41]. The major fields of interest addressed by these tools include comparative genomics which involves comparing the genetic content of one organism against that of another; prediction of the function of genes and sequences of the coding regions; identification of evolutionary events and inference of phylogenetic relationships. These fields of study play a critical role in elucidating pathogen evolution, niche adaptation, population structure and host-pathogen interaction. Furthermore, these findings inform vaccine and drug design, as well as the identification of virulence genes.

4. Bioinformatics pipelines and workflows

Bioinformatics pipelines and workflows comprise of a series of third-party executable command line software assembled to perform a specific task or analysis. A complete pipeline will, therefore, be able to support the end of analysis of a given field of study such as phylogenetics or variant detection. Pipelines can thus be broken down into two major components i.e. the data processing component and the analytical component that performs the core analysis of the pipeline. Below, we review some of the prominent bioinformatics pipelines and workflows that support the processing and analysis of NGS data to provide insights on relevant global surveillance of arboviral outbreaks.

5. Virus discovery and identification tools

Viral discovery and identification from isolates and metagenomic samples present major challenges to bioinformatics in general. This is because viral genomes are prone to very high variability and deviation from reference genomes [42], continuous emergence of new viruses with no available references, high intrapopulation diversity, and the relative rareness of viral DNA fragments in metagenomic samples [43]. These challenges have largely been addressed through the following pipelines.

5.1 Genome detective

Genome detective (http://www.genomedetective.com/app/) is an easy to use web-based software application that assembles the genomes of viruses quickly and accurately, designed to generate and analyze whole or partial viral genomes directly from NGS reads within minutes [44]. The application gains accuracy by using a novel alignment method that uses a combination of both amino acids and nucleotide scores to construct genomes by reference-based linking of de novo contigs. Speed and accuracy are also gained by using DIAMOND [45] with a UniProt90 reference dataset to sort viral taxonomy units. The use of DIAMOND and UniRef90 allowed genome detective to identify viral short reads at least 1000 times faster than when Blastn and the viral nucleotide database of NCBI were used. The software was optimized using synthetic datasets to represent the great diversity of virus genomes. The application was then validated with next-generation sequencing data of hundreds of viruses.

5.2 VirusTAP: viral genome-targeted assembly pipeline

One of the major difficulties in this process is the correct de novo assembly of viral genomes from crude metagenomic deep sequencing reads, including large amounts of bacteria and human related sequencing reads. Such read contaminations often force the server to overload during de novo assembly and might cause misassembly of the resultant contigs. Pre-filtering by host-mapping subtraction could lead to efficient de novo assembly, allowing the rapid and accurate procurement of a complete viral genome sequence. In addition to the accuracy of de novo assembly, the exclusion of human-related sequences can circumvent conflicting ethical issues by avoiding analyzing the personal genetic information of patients [46, 47].

VirusTAP is web-based, integrated NGS analysis tool designed to facilitate rapid and accurate viral genome assembly from raw reads by just clicking on several selections. Like genome detective, it ensures that non-viral reads are eliminated prior to de novo assembly in order to ensure performance is not compromised.

5.3 Virus identification pipeline (VIP)

VIP (https://github.com/keylabivdc/VIP) is a web-based virus discovery and identification tool [46]. With a single click, it will filter out background-related reads, classify reads on basis of nucleotide and remote amino acid homology, and perform phylogenetic analysis to provide evolutionary insights.

5.4 TAR-VIR: a pipeline for TARgeted VIRal strain reconstruction from metagenomic data

TAR-VIR is a non-reference based NGS analysis tool for the reconstruction of viral strains from metagenomic samples [46, 47]. It was developed to classify RNA viral reads from viral metagenomic data and also to produce the assembled viral strains (i.e. haplotypes) from classified reads. It mainly has two components: (1) viral read classification using partial or remotely related reference genomes and (2) de novo assembly of viral haplotypes from recruited reads with PEHaplo [47, 48], which is a haplotype reconstruction tool. As TAR-VIR has a modular structure, the users have options to use other assembly tools after read classification in step (1).

6. Genotyping tools

While variant discovery and identification tools play a critical role in determining the pathogen responsible for the infection, they are unable to determine the subtype or quasispecies that is responsible for the outbreak. Arboviruses exist as a mixed population of genomic variants due to rapid replication and the error prone nature of viral RNA-dependent RNA polymerase (RdRp) [47]. Monitoring virus genotype diversity is therefore crucial to understand the emergence and spread of outbreaks. Genotyping tools provide an efficient workflow to enable researchers and public health practitioners to determine the strain that is responsible for the outbreak.

Most free-access bioinformatics programs used to classify the genetic profile of subtypes, genotypes, subgroups or groups of viruses are based on the use of similarity search tools to determine the genotype of a new sequence. These genotyping tools use a set of reference sequence genomes, carefully selected for the purpose of representing each individual genotype. The use of a number of reference sequences representing the genotype of a given group increases the consistency and reproducibility of the data, thus ensuring a higher speed in the search for the data and offering greater and more complete information while ensuring that the results are not limited to an inadequate set of reference sequences that do not represent the information needed to identify the virus.

The similarity-based methods are useful for identifying recombination patterns in viral sequences, but they need further confirmation of their own phylogenetic methods and have no statistical support for their results.

Recently [49], four viral genotyping tools for yellow fever (YFV) (https://www.genomedetective.com/app/typingtool/yellowfevervirus/), dengue (DENV) (https://www.genomedetective.com/app/typingtool/dengue/), Chikungunya (CHIKV) (https://www.genomedetective.com/app/typingtool/chikungunya/) and Zika (ZIKV) (https://www.genomedetective.com/app/typingtool/zika/) were developed and linked to genome detective to enable phylogenetic classification below species level [50, 51].

6.1 Castor

The classification and annotation of virus genomes constitute important assets in the discovery of genomic variability, taxonomic characteristics and disease mechanisms. Existing classification methods are often designed for specific well-studied families of viruses [43]. Thus, the viral comparative genomic studies could benefit from more generic, fast and accurate tools for classifying and typing newly sequenced strains of diverse virus families.

CASTOR is a virus classification platform based on machine learning methods, inspired by a well-known technique in molecular biology: restriction fragment length polymorphism [52]. It simulates, in silico, the restriction digestion of genomic material by different enzymes into fragments. It uses two metrics to construct feature vectors for machine learning algorithms in the classification step. The performance of CASTOR, its genericity and robustness could permit the conduct of novel and accurate large-scale virus studies. The CASTOR web platform provides an open access, collaborative and reproducible machine learning classifiers. CASTOR can be accessed at (http://castor.bioinfo.uqam.ca).

7. Phylogenetic and phylodynamic tools

Phylogenetic tools are an extremely important resource used in the field of virology to study viral evolution, trace the origin of epidemics, establish the mode of transmission, investigate the occurrence of drug resistance or determine the origin of the virus in different body compartments. Thus, the tools developed by bioinformatics are fundamental to monitor the evolution of viral diversity, supporting studies of genomic sequence analysis, crucial for the surveillance of viral polymorphism, the development of new therapeutic strategies, the development of vaccine products or the appropriate choice products. Toward the development of a global surveillance outbreak surveillance system, the advances below have been made.

7.1 Nextstrain (https://nextstrain.org/)

Nextstrain is a real-time pathogen evolution tracking platform that implements cutting-edge analysis and visualization of pathogen genome data [53]. It provides evolutionary information in the form of interactive visualizations to virologists, epidemiologists, public health officials and citizen scientists. It has been used to track various arboviral epidemics globally including West Nile Virus (WNV) in the Americas, Zika virus in 33 countries and Dengue virus outbreaks in 64 countries. The platform is continually updated with publicly available datasets to provide new insights into viral epidemic outbreaks globally in an intuitive and visually esthetic manner.

8. Functional prediction tools

In disease surveillance, understanding the effect of mutations detected in the viral genomes through the methods identified above is invaluable in the development of relevant controls and interventions [47]. Many of these mutations serve as drug targets as well as provide insights into the response mechanism of the pathogens to existing interventions. A global surveillance system would therefore be incomplete without the capability to provide insights to the function of discovered mutations. Below we explore some of the tools that have been applied to understand the functional relevance of mutations found in arboviruses.

8.1 The SIFT (sorting intolerant from tolerant)

The SIFT algorithm predicts the effect of coding variants on protein function [54, 55]. Since its introduction in 2001, SIFT has become one of the standard tools for characterizing missense variation. It has a corresponding website that provides users with predictions on their variants.

9. Conclusion

Augmenting epidemiological data with insights from genomic data provides a powerful tool for surveillance and control of disease outbreaks. Advances in bioinformatics particularly leverage large genomic datasets to determine pathogenic organisms responsible for the outbreak, the origin of the infection and mutations responsible unique phenotypic traits. This information is crucial for effective planning interventions and combating outbreaks. An area of research interest that remains to be explored is the development of online platforms to perform functional analyses of statistically significant mutations in arboviruses. This information is invaluable in the development of vaccines and identification of drug targets.

Acknowledgements

This work was supported by the ZiBRA2 project supported by the Brazilian Ministry of Health (SVS-MS) and the Pan American Organization (OPAS) and founded by Decit/SCTIE/MoH and CNPq (440685/2016-8 and 440856/2016-7); by CAPES (88887.130716/2016-2100, 88881.130825/2016-2100 and 88887.130823/2016-2100). MG is supported by Fundação de Amparo à Pesquisa do Estado do Rio de Janeiro—FAPERJ.

Conflict of interest

The authors declare no conflict of interest.

Appendices and nomenclature

RT-qPCR	real time quantitative polymerase chain reaction
NGS	next generation sequencing
DNA	deoxyribonucleic acid
RNA	ribonucleic acid
VIP	virus identification pipeline
TAR-VIR	targeted viral
RdRp	RNA-dependent RNA polymerase
YFV	yellow fever virus
DENV	dengue virus
CHIKV	Chikungunya virus
ZIKV	Zika virus
WNV	West Nile virus
SIFT	sorting intolerant from tolerant

DOI: http://dx.doi.org/10.5772/intechopen.88769

Author details

Vagner Fonseca[1,2], Joilson Xavier[2], San Emmanuel James[1], Tulio de Oliveira[1], Ana Maria Bispo de Filippis[3], Luiz Carlos Junior Alcantara[2,3] and Marta Giovanetti[2,3]*

1 KwaZulu-Natal Research Innovation and Sequencing Platform (KRISP), College of Health Sciences, University of KwaZuluNatal, Durban, South Africa

2 Laboratório de Genética Celular e Molecular, ICB, Universidade Federal de Minas Gerais, Belo Horizonte, Minas Gerais, Brazil

3 Laboratório de Flavivírus, Instituto Oswaldo Cruz Fiocruz, Rio de Janeiro, Brazil

*Address all correspondence to: giovanetti.marta@gmail.com

IntechOpen

© 2019 The Author(s). Licensee IntechOpen. This chapter is distributed under the terms of the Creative Commons Attribution License (http://creativecommons.org/licenses/by/3.0), which permits unrestricted use, distribution, and reproduction in any medium, provided the original work is properly cited.

References

[1] WHO. Vector-Borne Diseases. 2017. Available from: https://www.who.int/news-room/fact-sheets/detail/vector-borne-diseases

[2] World Economic Forum. Global Risks 2015. World Economic Forum. Insight Report. 10th ed2015. Available from: http://www3.weforum.org/docs/WEF_Global_Risks_2015_Report15.Pdf

[3] LaBeaud AD. Why arboviruses can be neglected tropical diseases. PLoS Neglected Tropical Diseases. 2008;**25**:6-247. DOI: 10.1371/journal.pntd.0000247

[4] Powell JR. Mosquito-borne human viral diseases: Why Aedes aegypti? The American Journal of Tropical Medicine and Hygiene. 2018;**98**:1563-1565. DOI: 10.4269/ajtmh.17-0866

[5] Leta S, Beyene TJ, De Clercq EM, Amenu K, Kraemer MU, Revie CW. Global risk mapping for major diseases transmitted by *Aedes aegypti* and *Aedes albopictus*. International Journal of Infectious Diseases. 2018;**67**:25-35. DOI: 10.1016/j.ijid.2017.11.026

[6] Bhatt S, Gething PW, Brady OJ, Messina JP, Farlow AW, Moyes CL, et al. The global distribution and burden of dengue. Nature. 2013;**496**:504-544. DOI: 10.1038/nature12060.

[7] PAHO. Dengue and Severe Dengue, Cases and Deaths for Subregions of the Americas. 2019. Available from: http://www.paho.org/data/index.php/en/mnu-topics/indicadores-dengue-en/dengue-regional-en/261-dengue-reg-ano-en.html

[8] PAHO. 2019. Chikungunya Total Cases. Available from: http://www.paho.org/data/index.php/en/mnu-topics/chikv-en/551-chikv-subregions-en.html

[9] PAHO. Zika Total Cases. 2019. Available from: http://www.paho.org/data/index.php/en/mnu-topics/zika.html

[10] WHO. Neglected Tropical Diseases in the Eastern Mediterranean Region. 2019. Available from: https://apps.who.int/iris/bitstream/handle/10665/275463/Fact_Sheet_CDT_2018_EN_20491.pdf?ua=1

[11] WHO. Dengue and Severe Dengue. 2019. Available from: https://www.who.int/news-room/fact-sheets/detail/dengue-and-severe-dengue

[12] WHO. Emergencies Preparedness, Response. Chikungunya. 2018. Available from: https://www.who.int/csr/don/archive/disease/chikungunya/en/

[13] Schar DL, Yamey GM, Machalaba CC, Karesh WB. A framework for stimulating economic investments to prevent emerging diseases. Bulletin of the World Health Organization. 2018;**96**:138. DOI: 10.2471/BLT.17.199547

[14] Messina JP, Brady OJ, Golding N, Kraemer MU, Wint GW, Ray SE, et al. The current and future global distribution and population at risk of dengue. Nature Microbiology. 2019;**1**:10-11. DOI: 10.1038/s41564-019-0476-8

[15] Franklinos LH, Jones KE, Redding DW, Abubakar I. The effect of global change on mosquito-borne disease. The Lancet Infectious Diseases. 2019;**01**:18-124. DOI: 10.1525/abt.2017.79.3.169

[16] Zanotto PMA, Leite LCC. The challenges imposed by dengue, Zika, and chikungunya to Brazil. Frontiers in Immunology. 2018;**9**:1960-1964. DOI: 10.3389/fimmu.2018.01964

[17] Machado AA, Estevan AO, Sales A, da Silva Brabes KC, Croda J, Negrão FJ.

Direct costs of dengue hospitalization in Brazil: Public and private health care systems and use of WHO guidelines. PLoS Neglected Tropical Diseases. 2014;**4**:8-104. DOI: 10.1371/journal.pntd.0003104

[18] Bangert M, Latheef AT, Pant SD, Ahmed IN, Saleem S, Rafeeq FN, et al. Economic analysis of dengue prevention and case management in the Maldives. PLoS Neglected Tropical Diseases. 2018;**27**:12-96. DOI: 10.1371/journal.pntd.0006796

[19] Staples JE, Shankar MB, Sejvar JJ, Meltzer MI, Fischer M. Initial and long-term costs of patients hospitalized with West Nile virus disease. The American Journal of Tropical Medicine and Hygiene. 2014;**3**:402-409. DOI: 10.4269/ajtmh.13-0206

[20] Rossi G, Karki S, Smith RL, Brown WM, Ruiz MO. The spread of mosquito-borne viruses in modern times: A spatio-temporal analysis of dengue and chikungunya. Spatial and Spatio-temporal Epidemiology. 2018;**26**:113-125. DOI: 10.1016/j.sste.2018.06.002

[21] Gardy JL, Loman NJ. Towards a genomics-informed, real-time, global pathogen surveillance system. Nature Reviews Genetics. 2017;**1**:2-256. DOI: 10.1038/nrg.2017.88

[22] Grubaugh ND. Tracking virus outbreaks in the twenty-first century. Nature Microbiology. 2019;**4**:10-19. DOI: 10.1038/s41564-018-0296-2

[23] Morens DM, Folkers GK, Fauci AS. The challenge of emerging and re-emerging infectious diseases. Nature. 2004;**430**:242-249. DOI: 10.1038/nature02759

[24] Daszak P, Cunningham AA, Hyatt AD. Emerging infectious diseases of wildlife—Threats to biodiversity and human health. Science. 2000;**287**:443-449. DOI: 10.1126/science.287.5452.443

[25] Morse SS. Factors in the emergence of infectious diseases. Emerging Infectious Diseases. 1995;**1**:7-15. DOI: 10.3201/eid0101.950102

[26] Versalovic J, Lupski JR. Molecular detection and genotyping of pathogens: More accurate and rapid answers. Trends in Microbiology. 2002;**10**:15, 12377563-21

[27] Sabat AJ, Budimir A, Nashev D, Sá-Leão R, van Dijl JM, Laurent F. Overview of molecular typing methods for outbreak detection and epidemiological surveillance. Euro Surveillance. 2013;**18**:20-380. DOI: 10.2807/ese.18.04.20380-en

[28] Shendure J, Ji H. Next-generation DNA sequencing. Nature Biotechnology. 2008;**26**:1135-1145. DOI: 10.1038/nbt1486

[29] Haagmans BL, Andeweg AC, Osterhaus ADME. The application of genomics to emerging zoonotic viral diseases. PLoS Pathogens. 2009;**5**:100-557. DOI: 10.1371/journal.ppat.1000557

[30] McHardy AC, Adams B. The role of genomics in tracking the evolution of influenza A virus. PLoS Pathogens. 2009;**5**:10-56. DOI: 10.1371/journal.ppat.1000566

[31] Tang P, Gardy JL. Stopping outbreaks with real-time genomic epidemiology. Genome Medicine. 2014;**6**:1-104. DOI: 10.1186/s13073-014-0104-4

[32] Holmes EC. Viral evolution in the genomic age. PLoS Biology. 2007;**5**:2-78. DOI: 10.1371/journal.pbio.0050278

[33] Quick J, Grubaugh ND, Pullan ST, Claro IM, Smith AD, Gangavarapu K. Multiplex PCR method for MinION and illumina sequencing of Zika and other

virus genomes directly from clinical samples. Nature Protocols. 2017;**12**: 12-61. DOI: 10.1038/nprot.2017.066

[34] Thézé J, Li T, du Plessis L, Bouquet J, Kraemer MU, Somasekar S. Genomic epidemiology reconstructs the introduction and spread of Zika virus in Central America and Mexico. Cell Host & Microbe. 2018;**23**:855-864. DOI: 10.1016/j.chom.2018.04.017

[35] Loman NJ, Constantinidou C, Chan JZM, Halachev M, Sergeant M, Penn CW. High-throughput bacterial genome sequencing: An embarrassment of choice, a world of opportunity. Nature Reviews Microbiology. 2012;**10**:599-606. DOI: 10.1038/nrmicro2850

[36] AL-Dewik NI, Qoronfleh MW, et al. Advances in Public Health. 2019;**2**:44-76. DOI: 10.1155/2019/3807032

[37] Zhang J, Chiodini R, Badr A, Zhang G. The impact of next-generation sequencing on genomics. Journal of Genetics and Genomics. 2011;**38**:95-109. DOI: 10.1016/j.jgg.2011.02.003

[38] Koboldt DC, Steinberg KM, Larson DE, Wilson RK, Mardis ER. The next-generation sequencing revolution and its impact on genomics. Cell. 2013;**155**:27-38. DOI: 10.1016/j.cell.2013.09.006

[39] Elliott LT, Sharp K, Alfaro-Almagro F, Shi S, Miller KL, Douaud G, et al. Genome-wide association studies of brain imaging phenotypes in UK Biobank. Nature. 2018;**562**:210-216. DOI: 10.1038/s41586-018-0571-7

[40] Hamid JS, Hu P, Roslin NM, Ling V, Greenwood CM, Beyene J. Data integration in genetics and genomics: Methods and challenges. Human Genomics and Proteomics. 2009;**86**: 90-93. DOI: 10.4061/2009/869093

[41] Hwang B, Lee JH, Bang D. Single-cell RNA sequencing technologies and bioinformatics pipelines. Experimental & Molecular Medicine. 2018;**50**:8-96. DOI: 10.1038/s12276-018-0071-8

[42] Manso CF, Bibby DF, Mbisa JL. Efficient and unbiased metagenomic recovery of RNA virus genomes from human plasma samples. Scientific Reports. 2017;**7**:41-73. DOI: 10.1038/s41598-017-02239-5

[43] Rose R, Constantinides B, Tapinos A, et al. Challenges in the analysis of viral metagenomes. Virus Evolution. 2016;**2**:01-22. DOI: 10.1093/ve/vew022

[44] Vilsker M, Moosa Y, Nooij S, Fonseca V, Ghysens Y, Dumon K, et al. Genome detective: An automated system for virus identification from high-throughput sequencing data. Bioinformatics. 2018;**2**:23-98. DOI: 10.1093/bioinformatics/bty695

[45] Buchfink B, Xie C, Huson DH. Fast and sensitive protein alignment using DIAMOND. Nature Methods. 2014;**12**:20-59. DOI: 10.1038/nmeth.3176

[46] Remita AM, Halioui A, Diouara AAM, Daigle B, Kiani G, Diallo AB. A machine learning approach for viral genome classification. BMC Bioinformatics. 2017;**18**:2-08. DOI: 10.1186/s12859-017-1602-3

[47] Chen J, Huang J, Sun Y. TAR-VIR: A pipeline for TARgeted VIRal strain reconstruction from metagenomic data. BMC Bioinformatics. 2019;**20**:3-05. DOI: 10.1186/s12859-019-2878-2

[48] Chen J, Zhao Y, Sun Y. De novo haplotype reconstruction in viral quasispecies using paired-end read guided path finding. Bioinformatics. 2018;**34**:2927-2935. DOI: 10.1093/bioinformatics/bty202

[49] Faria NR. Genomic and epidemiological monitoring of yellow fever virus transmission potential. Science. 2018;**36**:894-899. DOI: 10.1126/science.aat7115

[50] Fonseca V, Libin PJK, Theys K. A computational method for the identification of dengue, Zika and Chikungunya virus species and genotypes. PLoS Neglected Tropical Diseases. 2019;**13**:7-231. DOI: 10.1371/journal.pntd.0007231

[51] Remita MA, Halioui A, Malick Diouara AA, Daigle B, Kiani G, Diallo AB. A machine learning approach for viral genome classification. BMC Bioinformatics. 2017;**18**:208

[52] Hadfield J, Megill C, Bell SM. Nextstrain: Real-time tracking of pathogen evolution. Bioinformatics. 2018;**34**:4121-4123. DOI: 10.1093/bioinformatics/bty407

[53] Alexander TC, Laura DK. Insights into arbovirus evolution and adaptation from experimental studies. Viruses. 2010;**12**:2594-2617. DOI: 10.3390/v2122594

[54] Li Y. VIP: An integrated pipeline for metagenomics of virus identification and discovery. Scientific Reports. 2016;**6**:23-774. DOI: 10.1038/srep23774

[55] Sim N-L, Kumar P, Hu J. SIFT web server: Predicting effects of amino acid substitutions on proteins. Nucleic Acids Research. 2012;**40**:452-457. DOI: 10.1093/nar/gks539

Section 2

Vector Biology and Disease Prevention

Chapter 4

New Ways to Tackle Malaria

Susanta Kumar Ghosh and Chaitali Ghosh

Abstract

Malaria is one of the oldest tropical diseases and still remains a focus of attention. Sub-Saharan African countries contribute 90% of the total malaria cases in the world. The World Health Organization (WHO) has advocated eliminating this disease by 2030 with the existing strategies and tools. Many initiatives are underway by several organizations, and 38 countries have achieved the elimination goal. The main backbone of the elimination process is smart surveillance followed by prompt public health responses. The control of the disease mainly relies on treatment of malaria positive cases with anti-malarials namely artemisinin-based combination therapy (ACT) for *Plasmodium falciparum*. In India, chloroquine is still effective against *P. vivax*. Use of 8-aminoquinolines primaquine and more recently tefenoquine warrants testing of G6PD deficiency status to avoid unnecessary hemolysis. Vector control operations mainly depend on the use of long-lasting insecticidal nets (LLINs) and indoor residual spray (IRS) with insecticides. The threat of resistance draws an open challenge in both treatment and vector management. New initiatives on surveillance, treatment, chemoprevention, and vector control using modern techniques of artificial intelligence, machine learning, genetic engineering, and digital approach of community engagement have great potential to accelerate the malaria elimination process.

Keywords: malaria, elimination, smart surveillance, treatment, vector management, community engagement

1. Introduction

Malaria is one of the oldest parasitic tropical diseases, and it takes a huge toll on human lives. It also causes great economic loss. Almost half of the population in the world is under the threat of malaria mostly in the tropical and sub-tropical countries. About 90% of the total malaria burden occurs in sub-Saharan African countries. Efforts to eradicate/or eliminate malaria began after the discovery of the role of mosquitoes in malaria transmission by Ronald Ross in 1897. In the beginning of the 20th century, most of the mosquito control operations were aimed at larval control using larvicidal oil, larvivorous fish and environmental management. These efforts made significant impacts on malaria control. Everything changed with the introduction of dichloro diphenyl trichloroethane (DDT) in the mid-1940s. Many European countries and the USA successfully eradicated malaria with the application of DDT and vector sanitation strategies, and improving general living standard [1].

Malaria eradication program in India haves had mixed success. After successful results from pilot studies on DDT, the National Malaria Eradication Program (NMEP) was launched in 1958 from the National Malaria Control Program (NMCP) in 1953. There was a huge success that resulted in almost complete malaria eradication in the mid-1960s with 0.1 million cases and no deaths. A kind of complacency led to a slow rise in malaria cases, and a total of 6.4 million cases were reported in 1976. This was due to the development of resistance to DDT by vector species, especially by *Anopheles culicifacies* s.l. and chloroquine resistance by *P. falciparum* malaria [2].

A Modified Plan of Operation (MPO) was launched aiming to treat each fever case suspected to be malarial infection with a presumptive dose of anti-malarial drug especially chloroquine. This provided some respite but the malaria cases remained at a static level with occasional regional outbreaks. In 1995, a revised guideline named Modified Action Plan (MAP) gave some lead which renamed as National Malaria Eradication Program (NMEP) in 1999. Subsequently this program was more disease centric and named as National Vector Borne Disease Control Program (NVBDCP) in 2002 [3].

In 2017, 0.84 million malaria cases with 174 related deaths were reported from India, while WHO estimated 9.6 million cases with 16,723 malaria-related deaths. This may be due to different methods of case estimation. From this state of current situation in India, malaria elimination has been envisaged with an aim to achieve it by 2030 [1, 4].

2. Malaria elimination initiatives

The global malaria elimination framework was launched in 2007, and a detailed Global Technical Strategy (GTS) was released in May 2015 aiming to eliminate malaria by 2030. The three recommendations to achieve this goal strongly emphasize strengthening of smart surveillance; prompt diagnosis and treatment; and enhance elimination process. The GTS thus focuses on 35 countries in which to eliminate malaria by 2030, and India is one of them [4].

India is one of the countries that have signed the National Framework for Malaria Elimination (NFME). The WHO estimated 219 million malaria cases with 435,000 related deaths in the world in 2017. This was higher than the previous years. The WHO Director-General has called an aggressive new approach `High Burden to High Impact' [1]. Of the 11 high malaria burden countries 10 are from Africa, but India is also under this category. Nearly half of the global malaria occurred in Nigeria (25%), the Democratic Republic of the Congo (11%), Mozambique (5%) and 4% each by India and Uganda. This means India needs a special attention. The NFME has been designed to ease the burden in most high burden Indian states especially Odisha, Madhya Pradesh, Chhattisgarh and Jharkhand. The major attention should be on strengthening the surveillance which is still poor in many states [1, 3].

3. Strengthening of ongoing surveillance

Surveillance is the main pillar in the malaria elimination process. In most situations, ongoing surveillance is not consistent with the national guidelines resulting in poor estimates of malaria burden. This needs to be converted into smart surveillance. In the digital era, all surveillance systems should follow the concept of the 'test–treat–track' strategy [5]. Android-based mobile apps can be applied for quick dispensation of surveillance data from the field to the local administrator for immediate action. This system at district-level management is implemented in many African countries. In this way, the time lag between diagnosis and treatment can be minimized [6]. Tracking the patient for completion and follow-up of the treatment has wider effects on the local cycle of malaria [1]. WHO has developed surveillance and data analysis dashboards using district health information software 2 (DHIS 2) [4]. Such digital-based data systems will make the surveillance system smart and efficient.

Strategy to change in surveillance is also an important step to accelerate the malaria elimination process. China adopted the '1-3-7' strategy that promoted the elimination process with zero indigenous malaria cases in 2017. This strategy envisions the strategic action from diagnosis to treatment within 3 days and public health responses to vector management within day 7 of the case detection. This also makes an easy platform for establishment of personal communication in the community [7]. Indonesia also adopted the '1-2-5' strategy for surveillance and response protocol in malaria elimination; on day 1 case management and notification; on day 2 case classification and foci investigation; and by day 5 foci response and elimination [8].

In southwestern coastal Mangaluru city, Karnataka state, India, malaria has been endemic over two decades. The local authority has implemented indigenously developed digital handheld tablets (TABs) for smart surveillance. These TABs have been allotted to each health worker after proper training. Now no manual data collection is used in the city. The link of the software was also provided to the local hospitals and diagnostic labs. The data can be accessed to the local administrators for taking action on the feedback received. Here the '1-3-7-14' strategy has been adopted where positive case is registered with start of treatment on day 1; completion of treatment by day 3; on day 7 vector control activities with follow-up smear check, and on day 14 follow-up smear check and completion of radical treatment for *P. vivax* cases with primaquine. In this way, the initial response of the anti-malarial drugs can be assessed. The best part of this system is that all the data can be retrieved, and the program can be monitored at all levels, assessing the opening and closing of each case. This system creates a great scope of community awareness through person-to-person communication [9]. In the dashboard of this system, algorithms of specific data can be incorporated for possible prediction of malaria outbreaks. Thus, the concept of 'predict–perform–protect' can be established using artificial intelligence (AI) and machine learning.

A recent study in Bangladesh has found that the movement of people can be tracked from the mobile phone network which can help prediction of outbreaks of diseases such as malaria. This enables the health authorities to take preventive measures in time [10].

4. Quick, efficient and point-of-care diagnosis

Malaria microscopy is still the best method and gold standard for malaria diagnosis. A microscopist normally examines 60 blood smears per day. This includes staining and data maintaining. Now expert microscopist can detect 20 to 50 parasites/μl blood that means a 0.001 to 0.005% level of parasitemia. This is not the cases with regular microscopists where the sensitivity is low. Routine in-house training on the line of continued medical education program can improve the efficiency of the microscopists [11].

Recent deployment of Rapid Detection Tests (RDTs) have changed the malaria diagnosis at large, but it has failed to detect when the level is <100 parasites/μl blood. This has become a nagging problem in detecting very low numbers of infected red blood cells and sub-microscopic parasites especially gametocytes in *P. falciparum* cases. This necessitated an alternate system of diagnosis. An indigenously developed handheld real-time micro-PCR based PDA (personal digital assistant) device that can detect parasites as low as 1.3/μl blood has been used successfully. This is a point-of-care device that performs tests onsite within 40 minutes. About 80 cases can be performed per day. The result of the cases can be shared through an internet system. Thus, case management becomes more efficient and effective [12]. Recently, a genome mining based identical multi-repeat sequences (IMRS) qPCR assay has been developed for diagnosis of malaria infection. This diagnostic method can detect very low level of parasitemia that cannot be detected by the routine 18S rRNA-based diagnostic system [13].

5. Tackling sub-microscopic and asymptomatic cases

In recent years, asymptomatic and sub-microscopic cases are reported from many endemic countries. In fact, these two aspects are non-synonymous. Sub-microscopic malaria cases present very low levels of parasitemia which generally missed in the routine microscopic examinations. Such cases may be symptomatic, and in most situations, these are asymptomatic cases. It has also been observed that in most high endemic areas asymptomatic cases with detectable levels of parasites do not show symptoms. This is because of a high immune status of the individual patients. It is suggested that such cases may be monitored under hospital supervision and clinical algorithms can be drawn to know more specific symptoms. Possibly such patients may show some kind of symptoms and may be on alternate days which are indicative of chronic malaria cases. Differential diagnosis of such cases becomes very difficult since they normally do not show any routine symptoms. However, experience clinicians can diagnose and successfully treat them with scheduled anti-malarials.

On the other hand, asymptomatic cases do not show presentable routine symptoms. Once proper diagnosis is confirmed treatment becomes very easy. In our experience, patients having malaria-like symptoms who could not be diagnosed with routine tests even with RDTs, had been treated for other diseases, mostly with antibiotics, but also with anti-tubercular therapy (ATT) for a long time even months. It has been observed that most antibiotics with quinoline molecules and ATT with rifampicin have anti-malarial properties. But these therapies cannot completely eliminate malarial parasites rather reduces the cure rate [14]. Such cases show sub-microscopic level of parasites. Normally these parasites do not show normal morphological features under microscopy. Only expert microscopists can identify such drug-affected parasites. In such cases, it may deem necessary first to stop all medicines and wait for the fever or fever-like symptoms to appear, then treat them with effective anti-malarials after expert microscopy. All these exercises should be done under medical supervision. The post response and relief from agony of such patients are remarkable.

Generally an important question is raised by most public health experts whether asymptomatic cases may cause potential risk of source of malaria transmission in endemic areas. In most endemic areas with high *P. falciparum* cases, residual load of gametocytes remain active in the blood circulation for a considerable period even after successful treatment, either with artemisinin-based combination therapy (ACT) or radical treatment with primaquine [1]. This can be solved with a simple *ex vivo* tests for detecting the presence of exflagellation. If this happens, it will indicate the potency of the gametocytes. This can be further extended to artificial membrane or direct feeding on patients (after obtaining human ethical approval) to *Anopheles stephensi* which will be kept at temperature of 28°C and 70–75% relative humidity (RH) in a controlled chamber for 7–8 days. Gut dissection on day 7 post-feeding will confirm the presence of oocysts. The presence of oocysts will indicate the potential threat of such gametocytes and their role in active transmission. It would be better to know the male and female gametocyte ratio before the experiment is performed. Generally 1 male to 3–5 female gametocytes sex ratio is found in *P. falciparum* [15]. Post-treatment gametocytemia is commonly detected by a quantitative nucleic acid sequence-based amplification (QT-NASBA) method [16]. It is better to kill all the fed mosquitoes on day 8 post-feeding so that there will be no threat of accidental release of infected mosquitoes for possible malaria infections.

6. Treatment of malaria cases

Chloroquine the cheapest anti-malarial drug is no longer prescribed for the treatment of *P. falciparum* malaria cases. Even monotherapy with artemisinin is also not recommended. Currently different ACTs are prescribed for treatment of *P. falciparum* malaria including some *P. vivax* cases. The partner drugs are sulfadoxine-pyrimethamine, piperaquine, lumefantrine, pyrrolidine, etc. In India, chloroquine is still efficacious against *P. vivax* even in the presence of *Pvcrt-0* and *Pvmdr-1* mutations [17]. On the other hand, the presence of *kelch 13* (*k13*) mutations in *P. falciparum* is linked to artemisinin resistance. Recent report from eastern India indicated the presence of two mutations G625R and R539T in 5/72 *P. falciparum* cases treated with artemisinin that linked to its presence of resistance [18]. In Africa, where *P. falciparum* is predominant there is no sign of artemisinin resistance even with more than 200 non-synonymous *k13* mutations recorded. A recent report suggested that artemisinin resistance in a patient can be addressed by changing the partner drugs which is responding to the local parasites [19]. Such combination should be selected for *P. falciparum* treatment. Several studies indicated the absence of S769 N mutation in *Pf*ATPase6 gene responsible for artemisinin resistance [20]. Possibly this marker would be the better one for monitoring of artemisinin resistance in most malaria-endemic settings.

6.1 Primaquine and tefenoquine for radical treatment

Primaquine – an 8-aminoquinoline is used for radical cure. In case of *P. vivax* 15 mg per day for 14 days and for *P. falciparum* a single dose of 45 mg is administered for eliminating hypnozoites for the former and gametocytes for the later, respectively. Recently tefenoquine, another 8-aminoquioline, has been recommended. But both of these drugs may cause possible hemolysis in G6PD-deficient patients [21]. Attempts are being made to find newer molecules to address this issue. In this regard, Medicines for Malaria Venture (MMV) is playing a primary role [22].

7. Change in vector control strategy

There are 465 *Anopheles* mosquitoes in the world, of which many members have sibling species complexes. Approximately 70 of them are capable for human malaria transmission [23]. Application of public health insecticides is the main strategy for vector control. DDT is the main insecticide and is partially responsible for most malaria elimination in Europe and Americas along with general improvement of living standards, and an effective detection and treatment program. Other countries missed out this opportunity to achieve this feat. Prolongation of its use lead to the development of resistance in the mid-1970s and also recorded the highest number of malaria cases. Other insecticides namely malathion (organophosphorous) and subsequently synthetic pyrethroids (deltamethrin, alphacypermethrin, lambdacyhalothrin, cyfluthrin, etc.) are used in the program. In some endemic areas, triple resistance has been recorded against the main rural vector *An. culicifacies*. Currently long lasting insecticidal nets (LLINs) impregnated with synthetic pyrethroids are widely used in the program. In general, behavioral changes of vector mosquitoes are a common phenomenon due to continuous use of IRS insecticides. This leads to outdoor resting and feeding behavior which are responsible for outdoor transmission. Some species change their biting time also, and thus becomes difficult in managing vector control operations [24].

7.1 Outdoor and residual transmission

Generally outdoor and residual transmissions are considered as the same phenomenon. But these are separate issues and would be dealt separately. Outdoor transmission occurs when local community engages on outdoor duties due to professional compulsion. This is most prevalent in forest fringe areas. For example, *An. dirus s.l.* is the most dominant species in Southeast Asia region. Here LLINs have very limited role. Many methods have been suggested, but none have been used for any practical purposes. Several traps have been developed in recent years; some are light-based, some CO_2- based, some octanol, commercial attractants based, and some with combination of all. Many experts recommend covering the whole body with proper clothing especially for security personnel, use of mosquito repellents and chemoprevention [1]. Recently Center for Disease control and Prevention (CDC) has approved the lemon eucalyptus oil for general use as mosquito repellent [25].

The most disturbing fact is that in most village settings human and cattle have mixed dwellings. This encourages the zoophagic mosquitoes to move from the bovine host to the humans. Here, a community-level action is needed. Experts recommend that all cattle dwellings should be located on the periphery of the village so there would be a spatial barrier between the foraging mosquitoes and humans. In this way, a strong zooprophylaxis would be established and direct human biting can be avoided [26].

7.2 Removing shrubs around houses

Outdoor transmission can be effectively contained when flowering shrubs around houses are removed. A study in Mali supports such concepts. The selected villages where flowering branches of invasive shrub *Prosopis juliflora* were removed experienced a 69.4% drop in *Anopheles* population density and a shifting of species composition [27].

7.3 Attractive toxic sugar bait (ATSB)

Like removing shrubs, ATSB is an alternate strategy to eliminate mosquitoes. Sugar bait of 10% sucrose mixed with 0.01% ivermectin soaked in sponge bait knocked down over 95% of *An. arabiensis* population [28]. But it requires community engagement for proper implementation [29].

Studies should be carried out to define the bionomics of local vectors. This will provide valuable information for planning proper vector control strategies. This should be an ongoing program. In the malaria elimination program, routine monitoring of vectors will allow appropriate decisions for effective control. Residual transmission is a resultant of presence of sub-microscopic level of malarial parasites in the community. This happens when intensive control measures overlook the residual presence of parasites. Such a situation happens when a type of complacency prevails and the surveillance system becomes fatigued. Many local-level focal outbreaks happen, and the public health response activities for vector control fail to decimate such foci [30]. It is, thus important to have a strict surveillance system in place to avoid such residual transmission and outbreaks.

7.4 Gene editing

The recent advances in genetic engineering technology of CRISPR/Cas9 (Clustered Regularly Interspaced Short Palindromic Repeats/Cas9), a system targets specific stretches of DNA and edit genomes at specific locations. This tool of gene editing/drive technology can revolutionize malaria elimination efforts by identifying and targeting the local vectors. The aim should be to create transgenic mosquitoes that will not be able to carry the malaria parasites. Some success to create transgenic species of the main malaria vector of African countries, *An. Gambiae,* has been achieved [31]. Gene-driven mosquitoes do not follow the Mendelian law of inheritance. This technology can be effectively applied to eliminate the invasive and endemic species to maintain the conservation of biodiversity [1].

7.5 Fungal application for vector control

Entomopathogenic transgenic fungus *Metarhizium pingshaense* expressing the spider neurotoxin hybrid (met-hybrid) killed 99% of the mosquito population in a controlled trial in Burkina Faso. The study was conducted in a trial village of 600 square meters area or `mosquito sphere'. The test mosquitoes were killed within 45 days. The researchers hope to find out a new tool to eliminate malaria when insecticide resistance is a major problem [32]. Before this field trial, extensive laboratory experiments were carried out on the two important malaria species *An. coluzzii* and *An. gambiae s.l* [33].

7.6 Bioenvironmental control of disease vectors

This is a holistic approach of vector control practiced in the beginning of the 20th century. The main aim of this approach is source reduction of larval breeding habitats. In other words, larval source management is the key strategy that mitigates challenges of larval control. Minor engineering, filling up of pools and puddles near human habitats, and biological control are some methods of this strategy. Since the mid-1980s the ICMR- National Institute of Malaria Research, New Delhi, India has made pioneering work on this front with great successes. In this strategy, health education and community engagement is an integral part [34].

7.7 Paratransgenesis

Paratransgenesis is a process by which the genetically modified symbionts from a target insect express molecules within the vector that show refraction to pathogens they transmit. This is a novel approach, now used for the control of malaria, trypanosomiasis and dengue. Recently, for the first time, we found *Veillonella* sp. in the gut of *An. stephensi* which may play an important role in paratransgenesis. A diverse microbial community was recorded in the salivary glands in *An. culicifacies*—the main malaria vector in rural India [35]. However, this strategy has to go a long way for the involvement of gene modification technology [36].

7.8 Jhum cultivation

Jhum or jhoom cultivation or slash and burn cultivation is a common practice of cultivation among tribal populations of northeast India and also in some hilly districts of Bangladesh. This practice of cultivation is linked with malaria transmission [37]. Besides using LLINs, it is important to find the main breeding habitats of vector species mainly *An. dirus* s.l. during the dry season. Control of such vector species can be obtained with the application of larvivorous fish mainly in wells which act as the ecological niche [1]. Further research may find a new way of intervention strategies.

7.9 Intervention with endectocide

Use of endectocide namely ivermectin in mass drug administration program is a potential intervention strategy to reduce residual transmission of malaria. This drug is used in elimination of human lymphatic filariasis and onchocerciasis programs. This old drug was developed from a natural substance by Satoshi Omura from Kitasato Institute, Japan and was further developed by William Campbell from Merck Lab originally for use in veterinary health program. But its use in onchocerciasis program was recognized for Nobel Prize for Physiology or Medicine in 2015. This is used as a potential tool in vector control program when anti-mosquito activities were recognized. Twenty-three projects under Malaria Elimination Science Alliance (MESA) are underway and their results will be available by 2020 that will be able to take a decision on the future use of endectocides in malaria control operations [38]. Besides the mentioned diseases, recent publication has given an overview on the use of ivermectin for various neglected tropical diseases (NTDs) that include ascariasis, trichuriasis, strongyloidiasis, loiasis (human *Loa loa*) and mansonelliasis [39]. However, adverse effects of ivermectin on local environment have been reported. Certain effects on dung beetles *Caccobius jessoensis, Copris ochus* and *Co. acutidens* have been reported in Japan [40].

7.10 Improved method of mosquito culture

It is important to grow fit and healthy mosquitoes under laboratory conditions for anyone working on them. Various methods of culturing of several species of mosquitoes are available. Most of malaria research is linked with several species *Anopheles* mosquitoes. Of them *An. stephensi* is widely used for its easy adaptation under laboratory conditions. Routine procedures were followed to colonize this species [41–43]. But some modifications in the routine methods gave a significant result in mass rearing of *An. stephensi* [44]:

i. Larval and adult rooms were maintained separately.

ii. For larval room RH was maintained at 45–50%, while temperature at 28°C.

iii. The adult room was maintained 12 hour dark and light periods; temperature at 28°C, and RH at 70–75%. Strict monitoring of temperature and humidity was maintained.

iv. The eggs laid by adult females in containers were bleached with freshly prepared 1% sodium hypochlorite solution for 1 minute under controlled pressure in a vacuum pump. For all purposes reverse osmosis (RO) water was used in the laboratory and pH of the water maintained around 7.00. About 250 bleached eggs were placed in especially designed white polypropylene trays (Polylab®, India, 375 × 300 × 75 mm^3) within a triangle made from disposable small straight straw pipes.

v. First instar larvae were emerged within 24 hours after bleaching. No food was given in the next 24 hours of hatching. Brewer's yeast powder dissolved in RO water was added in the tray water (300 ml) for the next 2 days.

vi. Subsequently a special larval food (Brewer's yeast and dog food at 70:30 ratio) were given daily for a specific amount depending on the stage of the larvae. Special scoop measuring 5–10 gm larval food in each tray depending on the larval age was provided. Pedigree brand dog food (chicken and vegetable mixed) were powered in a small grinder and mixed with Brewer's yeast. This formulation was different from earlier report. Pupae were visible from day 8 onwards.

vii. All pupae were harvested that developed up to day 12, and rest larvae were discarded following proper procedure. The pupae were bleached with freshly prepared 1% sodium hypochlorite solution for 1 minute, and placed in mosquito cages for emergence in to adults.

viii. The adult mosquitoes were provided with a mixture of 8% sucrose, 2% glucose mixed with 3% multivitamin kid syrup (Polybion® L, Merck Limited, India).

ix. Strict operational procedures were followed for maintaining sterile conditions for all steps.

x. This modified protocol would be very useful for mosquito research.

8. Discussion

The very decision to go for malaria elimination with the existing tools and intervention strategies was very challenging. Many initiatives have been undertaken. The President's Malaria Initiative (PMI) in 2005 to Malaria Elimination Research Alliance—India (MERA—India) in 2019 are all to accelerate the process of malaria elimination in all the high burden areas especially in Africa, Mekong Delta region and India, respectively [1]. A special initiative by the WHO in 2016, 21 endemic countries was identified for malaria elimination by 2020 i.e. E-2020 initiative [1]. In this direction zero malaria cases were reported from China and El Salvador in 2017. In 2018, Paraguay was certified as malaria free by the WHO. In 2019, Algeria achieved this goal. Three countries—the Islamic Republic of Iran, Malaysia and Timor-Leste—achieved zero malaria cases in 2018. In 2016, Sri Lanka achieved zero malaria certification, but in 2018 local transmission was reported from a case imported from India. But the local authorities immediately took action. Such quick public health response is required to maintain no transmission threat [45].

Vector control operations mainly rely on insecticide sprays. In most situations the spray operations are carried out by the local contract workers not properly trained; the spray equipments also not maintained properly; pressure not maintained while spraying; patchy and low coverage spraying; late supply of materials that force to defer the spray schedule; lack of supervision, low quality materials, improper storing warehouse, etc. All these confounding factors are responsible for continuation of transmission. Vector behavior also changes for prolonged insecticidal mode of operations [24].

Assessment of two important parameters—human blood index (HBI) and entomological inoculation rate (EIR) of important local vectors enable workers to develop an effective vector management. The global map of HBI of important malaria vectors revealed the highest index exists in African countries [26]. This indicates low ratio between human and animal populations forcing the vector mosquitoes feeding on human host. Emphasis on encouraging the local community to grow animals should be given priority which will change the transmission potential if the local vectors are primarily zoophagic [1].

Bioenvironmental control of vector populations is a part of the integrated vector management (IVM) concept. In this process control of other vectors of related diseases can also be achieved. *Swachh Bharat Abhiyan/Mission* (Clean India Movement) can be linked in such activity to gain better results. Larval Source Management using larvivorous fish is one of the intervention tools under IVM. This strategy is very effective when implemented at grass root levels with proper supervision and monitoring. Little or no serious emphasis has been given to this strategy. In many situations, it is overlooked and demeaned compared to other methods, mostly insecticides are available in hand. In fact, it works as a `social vaccine'. Globally, around 300 fish species have been identified as larvivorous nature. Two Poeciliid fish *Poecilia reticulata* and *Gambusia affinis* are widely used. The former is best for wells and other confined water bodies, while the later one is best for ponds and lakes [1]. There are many reports on adverse effects of the non-native fish on the local fish, but recent meta-analysis does not support this theory [46].

A new anti-larval product Aquatain AMF™ is available for anti-larval operation. It is a silicon-bases liquid (polydimethylsilicone—PDMS) formulation that forms a very thin film on standing water surface causing physical cover over its entire extension. The mosquito larvae are killed due to physical and mechanical action. There does not seem to develop insect resistance to this technique [47].

Community engagement through health education and empowering local policymakers help in taking appropriate decisions in vector control. Engaging some local school children as volunteers will laterally support such program [48]. In India, every year June is observed as anti-malaria month. Several activities highlighting the program on malaria are displayed. Local administration also actively takes part and makes some decisive actions.

Two vaccine candidates RTS, S/AS01 (TRADE NAME Mosquirix) and PfSPZ are under trial even though their protection level is moderately low. The former is undergoing phase 3 trial in children in three African countries. Possibly this will help reducing child mortality which is a major concern in most of African countries [1].

India contributes most of *P. vivax* malaria cases outside the high burden districts where *P. falciparum* is most dominant. Recent studies also indicate the presence of *P. ovale curtisi* and *P. ovale wallikeri*. *P. malariae* is also co-existent in the high burden tribal districts. All these are possible due to application of molecular diagnostic techniques [49]. Expert recommended genome mining studies which will unravel some underlying issues of malaria epidemiology [13]. Moreover, molecular DNA bar coding of parasites will also help in identifying their actual geographic origin [50]. This will surely assist in managing the drug resistance problem.

The recent identification of *P. knowlesi* as a human malaria with a zoonotic source prevalent in Malaysia and Southeast Asian countries may be very important [51]. Another disturbing factor is recent discovery of *P. falciparum* infection in two common Indian non-human primates *Macaca mulatta* and *M. radiata* [52]. This matter should be taken seriously when malaria elimination is underway. A distinct barrier between human and animal is essential to the ongoing elimination efforts.

9. Conclusion

There is a great movement and opportunity for malaria elimination globally. Many countries have already achieved this goal. The most success part of this movement is reporting zero malaria cases in China in 2017 and 2018, and preparing for malaria elimination certification in 2020. China made elaborate arrangements with full financial, administrative and operational commitment. This indicates that malaria elimination is possible with the existing tools and strategies. In the present situation dependence of insecticide should be minimized and promote other alternate strategies to avert the issue of insecticide resistance. India is making all efforts for a successful mission. Reduction of malaria cases and related deaths in 2017 is an indication. This was mainly the efforts made in eight most high burden districts in Odisha with the implementation of a program called *Durgama Anchalare Malaria Nirakaran* (DAMaN malaria elimination in inaccessible areas). Such initiatives should be implemented in other areas also [1].

In the elimination phase, there is a need to strengthen the existing public health system. The local health system should be quick and responsive to any malaria-related fevers. Routine in-house training, workshops should be conducted to maintain the malaria elimination momentum. A recent 5 days WHO workshop for South-East Asia Region (SEAR) recommended the global vector control response (GVCR). Entomologists from 11 countries participated in this workshop. Detailed reviews were exercised to find out the ongoing program implementation in each country. Such state-level workshops would help in malaria elimination and also other vector borne diseases [53].

Acknowledgements

The authors acknowledge the support from Indian Council of Medical Research, New Delhi; the Directors of ICMR-National Institute of Malaria Research, staff of ICMR- National Institute of Malaria Research, Bengaluru Field Unit. Also acknowledge the people in the community who helped and encouraged our studies from time to time.

Conflict of interest

The authors declare no conflict of interest.

Author details

Susanta Kumar Ghosh[1*] and Chaitali Ghosh[2]

1 ICMR-National Institute of Malaria Research, Bengaluru, India

2 Tata Institute for Genetics and Society, Bengaluru, India

*Address all correspondence to: ghoshnimr@gmail.com

IntechOpen

© 2019 The Author(s). Licensee IntechOpen. This chapter is distributed under the terms of the Creative Commons Attribution License (http://creativecommons.org/licenses/by/3.0), which permits unrestricted use, distribution, and reproduction in any medium, provided the original work is properly cited. CC BY

References

[1] Ghosh SK, Rahi M. Malaria elimination in India—The way forward. Journal of Vector Borne Diseases. 2019;**56**:32-40

[2] Dash AP, Valecha N, Anvikar AR, Umar A. Malaria in India: Challenges and opportunities. Journal of Biosciences. 2008;**33**:583-592

[3] National Vector Borne Disease Control Program. 2017. Available from: http://nvbdcp.gov.in/Doc/malaria-situation.pdf

[4] WHO. World Malaria Report 2017. World Health Organization; 2018

[5] Pradhan MM, Anvikar AR, Grewal Daumerie P, Pradhan S, Dutta A, Shah NK, et al. Comprehensive case management of malaria: Operational research informing policy. Journal of Vector Borne Diseases. 2019;**56**:56-59

[6] Quran V, Hulth A, Kok G, Blumberg L. Timelier notification and action with mobile phones–towards malaria elimination in South Africa. Malaria Journal. 2014;**13**:151

[7] Cao J, Sturrock HJW, Cotter C, Zhou S, Zhou H. Communicating and monitoring surveillance and response activities for malaria elimination: China's "1-3-7" strategy. PLoS Medicine. 2014;**11**(5):e1001642

[8] Sitohang V, Sariwati E, Fajariyani SB, Hwang D, Kurnia B, Hapsari RK, et al. Malaria elimination in Indonesia: Halfway there. Lancet Infectious Diseases. 2018;**6**(6):e604-e606. DOI: 10.1016/S2214-109X(18)30198-0

[9] Baliga S, Koduvattat N, Kumar M, Rathi P, Jain A. GIS based software technology assistance for effective control of malaria in Mangaluru, India. International Journal of Infectious Diseases. 2018;**73S**:222

[10] Stopping malaria outbreaks before they start. 2019. Available from: http://www.bbc.com/news/health-48581317

[11] Saha S, Narang R, Deshmukh P, Pote K, Anvikar A, Narang P. Diagnostic efficacy of microscopy, rapid diagnostic test and polymerase chain reaction for malaria using Bayesian latent class analysis. Indian Journal of Medical Microbiology. 2017;**35**:376-380

[12] Nair CB, Jagannath J, Pradeep AS, Prakash BN, Manoj NM, Malpani S, et al. Differential diagnosis of malaria on Truelab Uno1, a portable, real-time, MicroPCR device for point-of-care applications. PLoS One. 2016;**11**(1):e0146961

[13] Raju LS, Kamath S, Shetty MC, Satpathi S, Mohanty AK, Ghosh SK, et al. A genome mining based identification of identical multi-repeat sequences (IMRS) in plasmodium falciparum genome for highly sensitive qPCR assay and its application in malaria diagnosis. Journal of Molecular Diagnostics. 2019;**21**(5):824-838

[14] Pukrittayakamee S, Prakongpan S, Wanwimolruk S, Clemens R, Looareesuwan S, White NJ. Adverse effect of Rifampin on quinine efficacy in uncomplicated falciparum malaria. Antimicrobial Agents and Chemotherapy. 2003;**47**:1509-1513

[15] Robert V, Sokhna CS, Rogier C, Ariey F, Trape JF. Sex ratio of plasmodium falciparum gametocytes in inhabitants of Dielmo, Senegal. Parasitology. 2003;**127**:1-8

[16] Pett H, Gonçalves BP, Dicko A, Nébié I, Tiono AB, Lanke K, et al. Comparison of molecular quantification of plasmodium falciparum gametocytes by Pfs25 qRT-PCR and QT-NASBA in relation to mosquito infectivity. Malaria Journal. 2016;**15**:539

[17] Joy S, Mukhi B, Ghosh SK, Achur RN, Gowda DC, Surolia N. Drug resistance genes: Pvcrt-o and pvmdr-1 polymorphism in patients from malaria endemic South Western coastal region of India. Malaria Journal. 2018;**17**:40

[18] Das S, Saha B, Hati AK, Roy S. Evidence of artemisinin-resistant plasmodium falciparum malaria in eastern India. New England Journal of Medicine. 2018;**379**:1962-1964

[19] Wang J, Xu C, Fu Long Liao FL, Jiang T, Krishna S, Tu Y. A temporizing solution to "artemisinin resistance". New England Journal of Medicine. 2019;**380**:2087-2089

[20] Saha P, Naskar A, Ganguly S, Das S, Guha SK, Biswas A, et al. Therapeutic efficacy of artemisinin combination therapies and prevalence of S769N mutation in PfATPase6 gene of plasmodium falciparum in Kolkata, India. Asian Pacific Journal of Tropical Medicine. 2013;**6**:443-448

[21] White NJ. Tefenoquine – A radical improvement. New England Journal of Medicine. 2019;**380**:285-286

[22] Samby K, Ramachandruni H, Banerji J, Burrows JN, Grewal Daumerie P, Rob AM, et al. Partnering to fight malaria in India: Past, present and future. Journal of Vector Borne Diseases. 2019;**56**:15-24

[23] Sinka ME, Bangs MJ, Manguin S, Rubio-Palis Y, Chareonviriyaphap T, Coetzee M, et al. A global map of dominant malaria vectors. Parasites & Vectors. 2012;**5**:69

[24] Durnez L, Coosemans MM. Anopheles mosquitoes – New insights into malaria vectors. In: Manguin S, editor. Chapter 21: Residual Transmission of Malaria: An Old Issue for New Approaches. 2013. pp. 671-704. ISBN 978-953-51-1188-7

[25] CDC Confirms oil of lemon eucalyptus as effective as DEET. 2019. Available from: http://www.treehugger.com/lawn-garden/cdc-confirms-lemon-eucalyptus-oil-as-effective-as-deet.html

[26] Killeen GF. Characterizing, controlling and eliminating residual malaria transmission. Malaria Journal. 2014;**13**:330

[27] Muller GC, Junnila A, Tarore MM, Tarore SF, Doumbia S, Sissoko F, et al. The invasive shrub Prosopis juliflora enhances the malaria parasite transmission capacity of anopheles mosquitoes: A habitat manipulation experiment. Malaria Journal. 2017;**16**:237

[28] Tenywa FC, Kambagha A, Saddler A, Maia MF. The development of an ivermectin-based attractive toxic sugar bait (ATSB) to target anopheles arabiensis. Malaria Journal. 2017;**16**:338

[29] Maia MF, Tenywa FC, Nelson H, Kambagha A, Ashura A, Bakari I, et al. Attractive toxic sugar baits for controlling mosquitoes: A qualitative study in Bagamoyo, Tanzania. Malaria Journal. 2018;**17**:22

[30] Tiwari SN, Ghosh SK, Satyanarayan TS, Nanda N, Valecha N. Malaria outbreaks in villages in North Karnataka, India, and role of sibling species of anopheles culicifacies complex. Health. 2015;7:946-954

[31] James S, Collins FH, Welkhoff PA, Emerson C, Godfray EHC, Gottlieb M, et al. Pathway to deployment of gene drive mosquitoes as a potential biocontrol tool for elimination of malaria in sub-Saharan Africa: Recommendations of a scientific working group. American Journal of Tropical Medicine and Hygiene. 2018;**98**(Suppl 6):1-49

[32] Lovett B, Bilgo E, Millogo SA, Abel Kader Ouattarra AK, Sare I, Gnambani EJ,

et al. Transgenic Metarhizium rapidly kills mosquitoes in a malaria-endemic region of Burkina Faso. Science. 2019;**364**:894-897

[33] Bilgo E, Lovett B, Bayili K, Millogo AS, Sare I, Dabire RK, et al. Transgenic Metarhizium pingshaense synergistically ameliorates pyrethroid-resistance in wild-caught, malaria-vector mosquitoes. PLoS One. 2018;**13**(9):e0203529

[34] Dhiman R, Valecha N. Reducing malaria using environmental friendly approach – A Gandhian way of life. Indian Journal of Medical Research. 2019;**149**(Supplement):95-103

[35] Sharma P, Sharma S, Maurya RK, Das De T, Thomas T, Lata S, et al. Salivary glands harbor more diverse microbial communities than gut in anopheles culicifacies. Parasites & Vectors. 2014;7:235

[36] Coutinho-Abreu IV, Zhu KY, Ramalho-Ortigao M. Transgenesis and paratransgenesis to control insect-borne diseases: Current status and future challenges. Parasitology International. 2010;**59**:1-8

[37] Galagan SR, Prue CS, Khyang J, Khan WA, Ahmed S, Ram M, et al. The practice of Jhum cultivation and its relationship to plasmodium falciparum infection in the Chittagong Hill districts of Bangladesh. American Journal of Tropical Medicine and Hygiene. 2014;**91**:374-383

[38] Rabinovich NR. Ivermectin: Repurposing an old drug to complement malaria vector control. Lancet Infectious Diseases. 2018;**18**:584-585

[39] Hotez PJ, Alan Fenwick A, Molyneux DH. Collateral benefits of preventive chemotherapy — Expanding the war on neglected tropical diseases. New England Journal of Medicine. 2019;**380**:2389-2391

[40] Iwasa M, Maruo T, Ueda M, Yamashita N. Adverse effects of ivermectin on the dung beetles, Caccobius jessoensis Harold, and rare species, Copris ochus Motschulsky and Copris acutidens Motschulsky (Coleoptera: Scarabaeidae), in Japan. Bulletin of Entomological Research. 2007;**97**:619-625

[41] Devaiah MK, Pradeep AS, Sowmya KB, Ghosh SK, Sundaramurthy V, Sreehari U, et al. Influence of midgut microbiota in Anopheles stephensi on plasmodium berghei infections. Malaria Journal. 2018;**17**:385

[42] Kiattibutr K, Roobsoong W, Sriwichai P, Saeseu T, Rachaphaew N, Suansomjit C, et al. Infectivity of symptomatic and asymptomatic plasmodium vivax infections to a southeast Asian vector, Anopheles dirus. International Journal of Parasitology. 2017;**47**:163-170

[43] MR4 BEI Resources. Methods in Malaria Research. (2011). Available from: https://www.beiresources.org/portals/2/MR4/MR4_Publications/Methods%20in%20Anopheles%20Research%202014/2014MethodsinAnophelesResearchManualFull Versionv2tso.pdf

[44] Ghosh SK, Sowmya KB, Mukhi B, Varadharajan S, Sreehari U, Tiwari SN, et al. Plasmodium vivax platform in India. American Journal of Tropical Medicine and Hygiene. 2018;**99**:94

[45] Becoming malaria free by 2020. Available from: https://www.who.int/news-room/feature-stories/details/becoming-malaria-free-by-2020

[46] Walshe DP, Garner P, Adeel AA, Pyke GH, Burkot TR. Larvivorous fish for preventing malaria transmission. Cochrane Database of Systematic Reviews. 2017;(12):CD008090. DOI: 10.1002/14651858.CD008090.pub3

[47] AQUATAIN AMFTM Liquid Mosquito Film. 2019. Available from: http://www.aquatainexport.com/aquatain-amf

[48] Ghosh SK, Patil RR, Tiwari SN, Dash AP. A community-based health education for bioenvironmental control of malaria through folk theatre (Kalajatha) in rural India. Malaria Journal. 2016;**5**:123

[49] Krishna S, Bhandari S, Bharti PK, Basak S, Singh N. A rare case of quadruple malaria infection from the highly malaria-endemic area of Bastar, Chhattisgarh, India. PLoS Neglected Tropical Diseases. 2017;**11**(7):e0005558

[50] Tripathi M, Das A. Genotyping malaria parasites with DNA barcodes. Tropical Medicine and International Health. 2015;**20**:1636-1638

[51] Kantele A, Jokiranta TS. Review of cases with the emerging fifth human malaria parasite, plasmodium knowlesi. Clinical Infectious Diseases. 2011;**52**(11):1356-1362

[52] Dixit J, Zachariah A, Sajesh PK, Chandramohan B, Shanmuganatham V, Karanth KP. Reinvestigating the status of malaria parasite (plasmodium sp.) in Indian non-human primates. PLoS Neglected Tropical Diseases. 2018;**12**(12):e0006801

[53] Nagpal BN, Knox TB, Risintha P, Yadav RS, Ghosh SK, Uragayala S, et al. Strengthening of vector control in South-East Asia: Outcomes from a WHO regional workshop. Journal of Vector Borne Diseases. 2019;**55**:247-257

Chapter 5

Effect of Bendiocarb Indoor Residual Spraying on Entomological Inoculation Rate of *Anopheles arabiensis* in Northwestern Highlands of Ethiopia

Alemayehu Abate and Melaku Wale

Abstract

Entomological inoculation rate (EIR) is a method to estimate the level of human exposure to infective mosquito bites and assess impacts of vector control measures. The objective is to assess the effect of indoor residual spray (IRS) on blood meal index (BMI), sporozoite infection rate (SR), and EIR in *An. arabiensis* under local ecological settings in Ethiopia. A total 1541 fresh fed (FF) female *An. arabiensis* collected by CDC light trap and PSC were processed at the Center for Disease Control and Prevention Laboratory, Atlanta, Georgia, USA, to determine their BMI and SR, using enzyme-linked immunosorbent assay (ELISA). IRS reduced the abundance of FF female *An. arabiensis* in sprayed villages (n=62) while the number remained high in non-sprayed villages (n=1,690). The relative adjusted reduction in human blood feeding index (HBI) due to IRS varied between 3 and 10% except in 2014 when no human blood was detected in any of the three mosquitoes tested. The relative adjusted reduction in *P. falciparum* infection and EIR in *An. arabiensis* was 100% after IRS. The results illustrated that IRS was strong enough to reduce EIR in *An. arabiensis*. IRS is recommended to control malaria transmission in areas of similar ecological set.

Keywords: *A. arabiensis*, Ethiopia, EIR, IRS, vector control

1. Introduction

Current malaria vector control strategies rely heavily on indoor residual spraying (IRS) and long-lasting insecticide-treated mosquito nets (LLINs). The impact of these intervention tools on entomological malaria transmission risk factors needs to be evaluated. The level of exposure to infective mosquito bites could be measured using entomological inoculation rate (EIR) in the vector [1, 2]. The EIR is defined as the number of infective bites received by an individual per unit time (night, month, or year). It is the product of human-biting rate (HBR) and plasmodium sporozoite infection rate (SR) [3, 4].

The human landing catch (HLC) is the most commonly used method to determine the human-biting rate because it is the direct measure of human-vector contact [4]. However, due to ethical and logistic constraints associated with HLC, light trap catches (LTC), pyrethrum spray catches (PSC), and exit trap catches could be used as alternatives to human landing catches [3] to estimate the HBR. In this study, the Centers for Disease Control and Prevention (CDC) light trap and PSC mosquito sampling methods were used to estimate the HBR.

Malaria is a public health problem in Ethiopia. Indoor residual spraying and LLNs are the frontline pillars of malaria vector intervention tools that have been used in all malarious parts of the country. However, studies on the impact of these interventions on EIR are either limited or unavailable [5, 6]. Besides, EIR varies from region to region, even from locality to locality. Therefore, narrowing this knowledge gap would be valuable for vector control program. The present study was carried out to assess the impact of the current vector control strategy specifically IRS on BMI, SR, and EIR.

Dichlorodiphenyltrichloroethane (DDT) was the choice of insecticide for IRS operation that had been used for decades in many malarious areas of Ethiopia except at a few places where malathion was used for DDT-resistant vector populations. This was continued until 2007 when DDT was replaced by deltamethrin due to the development of DDT resistance in the major malaria vector populations [7]. Payable to the occurrences of deltamethrin resistance in different vector populations, in view of the possibility of cross-resistance between DDT and pyrethroid insecticides and the scaling up of the distributions of pyrethroid-treated LLINs, IRS control program again replaced deltamethrin by bendiocarb (carbamate group) in 2010 and still in use for IRS operations in different parts of the country.

The residual efficacy of bendiocarb with the recommended concentration could last between 2 and 6 months depending on the nature of sprayable surfaces [8]. Therefore, bendiocarb was the choice of insecticide used for IRS operation during the present study.

2. Materials and methods

2.1 Study site

The study was carried out in two adjacent villages, namely, Andassa (N11° 30′ 14.6″, 037° 29′ 27.8″) and Tikurit (11° 30′ 49.8″, 037° 28′ 02.8″), Bahir Dar Zuria District, North West Ethiopia. These villages were separated by Andassa River and buffered by about 2 km vegetable and fruit farms. The study villages were selected purposively by considering malaria endemicity and the history of IRS implementation. Indoor residual spraying and LLINs are the primary intervention tools that have been used for years against *A. arabiensis* (important vector of the study area). The vector has developed different levels of insecticide resistance to insecticides of different classes recommended for both LLIN treatment and IRS operation [9].

2.2 Design

A comparative study was carried out in Andassa and Tikurit villages. The study was conducted for 2 consecutive years. Andassa received two rounds of sprays one in 2013 and another in 2014, while no spray was implemented in Tikurit. Participants in the unsprayed villagers were provided with treated bed nets free of charge, and individuals found infected received free treatment at the nearest health center. The susceptibility status of *A. arabiensis* to bendiocarb was confirmed before the application of IRS.

2.3 Mosquito sampling

Adult female *A. arabiensis* were collected from 24 residential houses (12 houses/village and 6 houses/sampling method) using pyrethroid spray catch (PSC) and CDC light trap sampling methods. PSC was applied by spraying pyrethroid insecticide under which a white muslin cloth was placed to facilitate knocked-down mosquito collection. Human baits, sleeping in beds covered with treated bed nets, were used to reinforce CDC light traps. Mosquitoes that were collected by each sampling method before and after IRS were then stored individually in tubes containing silica gel to process and determine their BMI and SR in the lab.

2.4 Blood meal host source and sporozoite rate determination

Enzyme-linked immunosorbent assay (ELISA) originally described by Beier et al. [10] and CS-ELISA [11] protocols were adopted and used for BMI and SR analyses, respectively. Blood-fed mosquitoes preserved individually in tubes containing silica gel were used to determine their BMI and SR. Heads-thoraxes of mosquitoes were separated from their abdomens, and each body part (abdomen/head-thorax) was given a corresponding ID number and kept individually in tubes for analyses.

2.5 Blood meal source determination

The mosquito abdomen, which was kept individually in tubes containing silica gel, was ground in a tube containing 100 μl of phosphate buffer saline (PBS) with a plastic pestle fitted with foot-operated grinder. The pestle was rinsed twice with 200 μl of PBS to achieve the final volume of 500 μl. The samples were either incubated at room temperature for 3 h and then stored at 4°C and tested the next day. Mosquitoes were tested to assess the blood meal origin of human and bovine only because these hosts were the predominant hosts of the vector during the study period. A 96-well ELISA plate was used, and 50 μl of the positive control for the blood meal host being tested was loaded. Wells A2–A5 had 50 μl of the negative controls, and wells A6–A8 were blanks containing 50 μl of blocking buffer. The plate was then covered and incubated for 3 h. The mosquito triturate was then aspirated by multichannel pipet, and the plate was washed three times with 200 μl PBS-Tween20 (5%). For a full 96-well plate, the peroxidase conjugate anti-host IgG antibody was prepared by adding 4800 μl of blocking buffer and 19.2 μl of anti-host and 1 μl of 1:100,000 of each of the negative control [10]. Fifty microliter of peroxidase conjugate was added to each well, and the plate was covered and incubated for 1 h at room temperature. The plate was then washed three times with 200 μl PBS-Tween20 (5%), and the one component ABTS peroxidase substrate was added to each well. PBS-Tween20 was aspirated by multichannel pipet, and plates were banged between washes. After 30 min of covered incubation at room temperature, the plate was read with the SpectraMax 340 plate reader (Molecular Devices) at 414 nm.

2.6 Sporozoite rate determination

The head-thorax of a mosquito, which was kept individually in step tubes containing silica gel, was ground in 1.5 μl microcentrifuge grinding tube containing 50 μl PBS with a plastic pestle fitted with foot-operated grinder. The pestle was rinsed twice with 100 μl of PBS and dried with tissue paper to prevent contamination between mosquito samples.

A 96-well ELISA PVC plate was coated with 50 μl of capture monoclonal antibodies (mAb) of each plasmodium sporozoite species (Pf, Pv-2010, and Pv-247) in each well of the ELISA plates (a separate plate used for each species), covered and incubated for half an hour. After the well contents were aspirated, plates were banged upside down on paper towel five times. The wells were then filled with 200 μl blocking buffer (BB), covered with lid and incubated for 1 h at room temperature. Well contents are aspirated and the plate is banged on paper towel five times. Samples and controls were loaded into the plate (well 1A, positive control; wells 1B–1H, negative control; and the rest of the wells with mosquito triturate) and covered and incubated for 2 h. Well contents were aspirated, and the plates were banged upside down on paper towel five times and washed two times with 200 μl of PBS-Tween20. The wells were aspirated, and plates were banged upside down five times with each wash. Then a 50 μl of peroxidase conjugate solution of each plasmodium sporozoite species (Pf, Pv-2010, and Pv-247) was added to each well and covered and incubated for 1 h. After aspirating the well contents and banging the plates, wells were washed three times with 200 μl of PBS-Tween20 and aspirated, and plates were banged five times with each wash. Finally, a 100 μl of the substrate solution was added per well, covered with cover plate and incubated for 30 min. The results were then read visually at the SpectraMax 340 plate reader (Molecular Devices) at 405–414 nm. All positive samples were retested for confirmation.

2.7 Determination of entomological inoculation rate

Plasmodium EIR of *A. arabiensis* was determined based on CDC light traps and PSC. The EIR was estimated from PSC samples as described by the World Health Organization [12] using the formula: number of fresh fed (FF) mosquitoes caught by PSC/ no. human occupants who spent the previous night in sprayed house) × (number of human fed mosquitoes/number of mosquitoes tested for human blood meal) × (number of sporozoite positive ELISAs/ number of mosquitoes tested, i.e., HBR × CSP rate. The human-biting rate was calculated by dividing the total number of freshly fed *A. arabiensis* caught in PSC by the total number of occupants who slept in the houses in the previous night of mosquito collection and multiplied by the HBI. The HBI was calculated as the proportion of *Anopheles* mosquitoes that fed on humans to the total *Anopheles* analyzed for blood meal origin [13, 14]. EIR from CDC light trap catches was estimated using the standard method, 1.605 (number of circumsporozoite-positive ELISA results from CDC light trap/ no. of mosquitoes tested) × (number of mosquitoes collected by CDC LT/ no. of CDC LT catches), and the alternative method, 1.605 (no. positive ELISA/no. catches) [15].

2.8 Data analyses

The relative adjusted reduction in human blood feeding index (HBI), sporozoite rate (SR), and the entomologic inoculation rate (EIR) of the vector after intervention was calculated using the formula [Ref]: $PR = 100 - \frac{C1T2}{C2T1} \times 100$, where C1 and

C2 and T1 and T2 describe the either the number of *A. arabiensis* or percentages of BMI, SP, or EIR in sprayed (T) and non-sprayed villages (C) before IRS (subscript 1) and after IRS (subscript 2). This formula takes into account that changes in the mosquito population and parasite prevalence are taking place at the same level and rate in both sprayed and non-sprayed villages, i.e., the reductions were adjusted for the background differences. This formula was used only when the denominators were non-zero.

2.9 Ethical clearance

Ethical permission for the study was obtained from the Ethiopian Public Health Institute and Amhara Regional Health Bureau. Verbal consent was also obtained from the owner of each house sampled for mosquitoes. The study did not involve human or animal subjects.

3. Results

3.1 Effect of IRS on the abundance of *A. arabiensis*

Table 1 shows the abundance and abdominal status of *A. arabiensis* collected before and after spray. The abundance and abdominal status of *A. arabiensis* varied by sampling method, spray status, study village, and year. Among 5425 *A. arabiensis*, 3111 of them were collected by CDC light traps and 2314 of them by pyrethrum spray catches (PSC). The number of semi-gravid and gravid *A. arabiensis* was smaller in CDC light trap catches than in PSC collections. The proportions of unfed *A. arabiensis* were higher in CDC light trap catches than in PSC collections. Fresh fed *A. arabiensis* was dominant in PSC collections (>75%), while <54% of them were FF in CDC light trap catches. The abundance of these FF mosquitoes was declined after IRS in sprayed villages (n = 62), while the number of FF remained high in non-sprayed villages (n = 1690). The abundance of unfed, gravid, and semi-gravid mosquitoes also decreased after spray.

3.2 Effect of IRS on HBI

Among 3451 FF *A. arabiensis* collected, 1574 (45.61%) of them were tested to determine their blood meal sources and sporozoite infection rate. The relative adjusted reduction in *A. arabiensis* human blood feeding index (HBI) due to IRS implementation varied from 3 to 10% except in 2014 when no human blood was detected in any of the three mosquitoes that were collected and tested. Despite IRS implementation reduced HBI, a non-negligible proportion of *A. arabiensis* still fed on humans (**Table 2**).

3.3 Effect of IRS on SR

The estimated sporozoite rate in *A. arabiensis* was low in both sprayed and non-sprayed villages especially after IRS implementation. As indicated by ELISA test, *P. falciparum* was more prevalent than *P. vivax* in both sprayed and non-sprayed villages. *Pv*-247 was the only subspecies detected during the study period. There was no any mixed infection in the vector in both study villages during the study period. Neither *P. falciparum* nor *P. vivax* was not detected in *A. arabiensis* collected from sprayed villages after the implementation of

Year	Village	Before spray					After spray					Adjusted reduction (%)
		CDC light trap collection										
		Row total	UF	FF	SG	G	Row total	UF	FF	SG	G	
2013	Sprayed	103	46	57	0	0	12	6	6	0	0	8.6
	Non-sprayed	599	356	240	0	3	811	341	468	0	2	
	Column total	702	402	297	0	3	823	347	474	0	2	
2014	Sprayed	139	56	83	0	0	67	18	48	0	1	5.7
	Non-sprayed	146	69	71	0	6	1234	583	650	0	1	
	Column total	285	125	154	0	6	1301	601	698	0	2	
			987 pyrethrum spray collection 45.69% 2124 = 3111 53.65% AIRS									
2013	Sprayed	176	13	151	10	2	6	1	5	0	0	4.2
	Non-sprayed	769	33	666	49	21	624	19	543	48	14	
	Column total	945	46	817	59	23	630	20	548	48	14	
2014	Sprayed	471	16	302	86	67	3	0	3	0	0	3.9
	Non-sprayed	228	25	129	33	41	37	6	29	0	2	
	Column total	699	41	431	119	108	40	6	32	0	2	
1644		75.91 670 = 2314 86.57% AIRS 5425										

Table 1.
Abundance and abdominal status of A. arabiensis collected by PSC and CDC light traps from sprayed and non-sprayed villages in Bahir Dar Zuria District, North West Ethiopia, in 2013 and 2014.

Before spray				After spray		
Year	Host	Sprayed (n)	Non-sprayed (n)	Sprayed (n)	Non-sprayed (n)	Adjusted reduction (%)
			CDC light trap collection			
2013	HBI	19.30 (57)	18.18 (176)	16.67 (6)	17.61 (176)	-10.83
	BBI	31.58 (57)	42.05 (176)	33.33 (6)	40.91 (176)	+8.48
	Mix	19.30 (57)	1.7 (176)	16.67(6)	0 (176)	
	Un	29.82 (57)	38.07 (176)	33.33 (6)	41.48 (176)	+2.58
2014	HBI	18.75 (80)	18.57 (70)	16.67 (48)	17.05 (176)	-3.17
	BBI	33.75 (80)	44.29 (70)	37.50 (48)	46.02 (176)	+6.93
	Mix	7.5 (80)	0 (70)	0 (48)	0 (176)	
	UN	40 (80)	31.14 (70)	45.83 (48)	36.93 (176)	-3.39
			Pyrethrum spray sheet collection			
2013	HBI	20.71 (140)	25 (176)	20 (5)	25 (176)	-3.43
	BBI	36.43 (140)	51.70 (176)	40 (5)	55.11 (176)	+5.16
	Mix	20 (140)	0 (176)	20 (5)	0 (176)	
	UN	22.8 (140)	23.30 (176)	20 (5)	21.02 (176)	-2.76
2014	HBI	19.89 (176)	18.75 (80)	0 (3)	17.24 (29)	100
	BBI	32.95 (176)	48.75 (80)	33.33 (3)	48.28 (29)	+2.14
	Mix	24.43 (176)	21.25 (80)	66.67 (3)	0 (29)	
	UN	22.73 (176)	30.00 (80)	0 (3)	34.48 (29)	0
		453	502-955	62	557-619	

HBI, human blood index; BBI, bovine blood index; UN, unknown hosts; n, number of mosquitoes tested for their blood meal origin.

Table 2.
Effect of bendiocarb IRS on blood meal sources (BMS) of A. arabiensis in sprayed and non-sprayed villages, Bahir Dar Zuria District, North West Ethiopia, in 2013 and 2014.

IRS. Similar results were observed for *Pv*-247 in non-sprayed villages except in 2013 when SR was 0.57% in *A. arabiensis* caught by CDC light trap. The relative adjusted reduction in *P. falciparum* infection in *A. arabiensis* in sprayed villages was 100% after IRS. A similar result was observed for *Pv*-247 EIR in 2013 in *A. arabiensis* collected by CDC light traps (**Table 3**).

3.4 Effect of IRS on EIR

The reduction in EIR after the implementation of IRS had similar trends with the reduction in SR because EIR is the product of SR and HBI. Compared with CDC light trap catches, EIR was high in PSC catches, i.e., *Pf*-EIR in *A. arabiensis* was 452 infective bites/night/house in PSC catches, while it was 32.2 infective bites/night/house in CDC light trap catches. *Pv*-247 EIR was 226 and 16 infective bites/night/house in *A. arabiensis* collected by PSC and CDC light traps, respectively. The relative adjusted reduction in *Pf*-EIR in *A. arabiensis* was 100% after the implementation of IRS. A similar result was observed for *Pv*-247 EIR in 2013 in *A. arabiensis* caught by CDC light traps (**Table 4**).

Before spray				After spray		
Year	**Parasite**	**Sprayed (n)**	**Non-sprayed (n)**	**Sprayed (n)**	**Non-sprayed (n)**	**Adjusted reduction (%)**
			CDC light trap collection			
2013	Pf	1.75 (57)	1.14 (176)	0 (6)	0.57 (176)	100
	Pv-247	1.75 (57)	0.57 (176)	0 (6)	0. 57 (176)	100
	Pv-210	0 (57)	0 (176)	0 (6)	0 (176)	
	Mixed	0 (57)	0 (176)	0 (6)	0 (176)	
2014	Pf	2.5 (80)	1.43 (70)	0 (48)	1.70 (176)	100
	Pv-247	0 (80)	0 (70)	0 (48)	0 (176)	
	Pv-210	0 (80)	0 (70)	0 (48)	0 (176)	
	Mixed	0 (80)	0 (70)	0 (48)	0 (176)	
			Pyrethrum spray sheet collection			
2013	Pf	1.43 (140)	1.14 (176)	0 (5)	1.14 (176)	100
	Pv-247	0.71 (140)	0.57 (176)	0 (5)	0 (176)	
	Pv-210	0 (140)	0 (176)	0 (5)	0 (176)	
	Mixed	0 (140)	0 (176)	0 (5)	0 (176)	
2014	Pf	1.70 (176)	1.25 (80)	0 (3)	0 (29)	
	Pv-247	0 (176)	0 (80)	0 (3)	0 (29)	
	Pv-210	0 (176)	0 (80)	0 (3)	0 (29)	
	Mixed	0 (176)	0 (80)	0 (3)	0 (29)	

Pf, Plasmodium falciparum; Pv-247, Plasmodium vivax 247; Pv-2010, Plasmodium vivax 2010; n, number of mosquitoes tested for CSP ELISA.

Table 3.
Sporozoite rate of A. arabiensis (based on LTC).

Before spray				After spray		
Year	**EIR**	**Sprayed**	**Non-sprayed**	**Sprayed**	**Non-sprayed**	**Adjusted reduction (%)**
			CDC light trap collection			
2013	Pf	16	32.2	0	4.47	100
	Pv-247	16	16.1	0	4.47	100
	Pv-210	0	0	0	0	
	Mixed	0	0	0	0	
2014	Pf	26.76	13.38	0	13.34	100
	Pv-247	0	0	0	0	
	Pv-210	0	0	0	0	
	Mixed	0	0	0	0	
			Pyrethrum spray sheet collection			
2013	Pf	101.58	452.01	0	151.62	100
	Pv-247	50.44	226	0	0	
	Pv-210	0	0	0	0	

Before spray				After spray		
Year	**EIR**	**Sprayed**	**Non-sprayed**	**Sprayed**	**Non-sprayed**	**Adjusted reduction (%)**
	Mixed	0	0	0	0	
2014	Pf	249.2	88.83	0	0	
	Pv-247	0	0	0	0	
	Pv-210	0	0	0	0	
	Mixed	0	0	0	0	

EIR, entomological inoculation rate; Pf, Plasmodium falciparum; Pv-247, Plasmodium vivax 247; Pv-2010, Plasmodium vivax 2010.

Table 4.
Estimated the effect of IRS on EIR of A. arabiensis based on CDC light trap and pyrethrum spray sheet collection from sprayed and non-sprayed villages in Bahir Dar Zuria District, North West Ethiopia, in 2013 and 2014.

4. Discussion

The aim of vector control using IRS and LLIN interventions is to reduce vectors' abundance, survival, contact with human, and feeding frequency [16]. Vector abundance is an important determinant of malaria transmission [13, 14], and thus factors that increase or decrease vector abundance could have an impact on the intensity of disease transmission. The present study demonstrated that IRS implementation brought about 4–9% reduction in the abundance of *A. arabiensis* signifying that the abundance of this vector could not be reduced to non-detectable level by the implementation of IRS. Previous similar studies in Ethiopia are either missing or unavailable to compare and contrast with the present study. However, studies from Zambia [17] validated that the effect of IRS on the density of *A. arabiensis* was not as strong as on *A. gambiae* s.s and *A. funestus* due to its exophilic and wide-ranging feeding behavior. Alegana et al. [18] also confirmed that IRS intervention reduced the density of *A. funestus* and *A. gambiae* s.l disproportionally, twice as high on *A. funestus* compared with *A. gambiae* s.l. Thus, malaria transmission through the bites of *A. arabiensis* could not be intercepted entirely by the application of IRS so that the impact of IRS should be complemented by and integrated with other vector control interventions. Blood meal source analyses indicated that *A. arabiensis* was found to have strong preferences to bovine and other hosts over human hosts. Similar results from other parts of the country were published in previous studies [19–21] where *A. arabiensis* demonstrated strong blood meal preferences of bovine over human hosts. Similar results were also reported from neighboring Eritrea [22] and Kenya [23]. Contrary to zoophilic, strong athrophilic tendency was observed in *A. arabiensis* in Zambia [24–26]. The potential reason for the differences observed in the anthrophilic tendency of *A. arabiensis* between East and South African countries would be justified by the differences in their ecological setups and the impact of these ecological differences on the ecology and behavior of *A. arabiensis* populations in these two sub-African regions. The application of IRS in the present study further reduced the anthropophily of the vector signifying that zooprophylaxis could be considered as a potential malaria vector control strategy in areas having similar ecological setups with the present study site. On the contrary, a considerable proportion of *A. arabiensis* still fed on human hosts suggesting that zooprophylaxis alone cannot intercept malaria transmission. Thus, zooprophylaxis would advance the effectiveness of malaria interventions if used in an integrated way with other vector control intervention measures.

Either data are unavailable or no previous attempts were made about the impact of IRS on SR in Ethiopia. However, studies from other African countries [27, 28] demonstrated that the implementation of IRS reduced SR to non-detectable level, which is consistent with the results of the present study. And these would substantiate the contribution of IRS implementation in reducing malaria transmission risks in general and SR in particular in the present study area and others having similar ecological setups.

In the present study, *P. falciparum* was more prevalent than *P. viva* in *A. arabiensis.* No *A. arabiensis* was found positive for either *P. falciparum* or *P. vivax* in sprayed villages after IRS. Although too few *A. arabiensis* were recorded in sprayed villages after IRS, it would have been necessary to process thousands of mosquitoes to find any of them were infected by malaria parasites. There was no any mixed infection detected. The proportion of plasmodium-infected *A. arabiensis* was also low in non-sprayed villages indicating that SR might be low in naturally occurring vector population. Contradictory results about the prevalence of *P. falciparum* and *P. vivax* in *A. arabiensis* have been reported from different parts of Ethiopia at different times. Massebo et al. [21] reported the dominance of *P. falciparum* over *P. vivax*, while [6] reported the dominance of *P. vivax* over *P. falciparum* in South West Ethiopia. Animute et al. [29] reported the dominance of *P. vivax* over *P. falciparum* in South Central Ethiopia. Differing from all these, [30] reported that no *A. arabiensis* was found positive either for *P. falciparum* or *P. vivax* in South West Ethiopia. Except Taye and his colleagues [30], other investigators used either CDC or PSC mosquito sampling method so that the differences observed in the prevalence of malaria parasites in *A. arabiensis* could be potentially justified by the differences in ecological setups of the study sites and time period in which the study was conducted. Otherwise, this would be a question of validation.

Malaria transmission intensity, which is normally expressed by EIR, is highly variable with annual EIRs ranging from < 1 to >1000 infective bites per person per year in Africa [31]. Variations in EIR in malaria vectors could be due to different factors such as ecological heterogeneity at continental, regional, and country level [29, 32, 33] and season (dry or wet) [29, 34, 35]. For example, the burden of malaria is high in tropical countries having warm temperature, heavy rainfall, high humidity, and efficient *Anopheles* vectors than nontropical countries [36]. Previous studies indicated that the impact of wet or dry season on EIR is inconsistent, i.e., published reports indicated that EIR is higher during wet season [15, 35, 37] or vice versa [38, 39].

In the present study, a very high *Pf*-EIR was observed in the vector in both years and study villages although SR and HBI were low. The trend was also similar for *Pv*-247 EIR in both study villages before IRS in 2013. These findings would be justified by the occurrences of high mosquito density during the study periods. The level of EIR of both parasites went to zero in sprayed villages after the implementation of IRS suggesting that IRS application is 100% effective to control disease transmission. In contrast, previous studies reported that EIR was 90% lower in the ITN community and 93% lower in the IRS community, relative to the community without intervention. The differences observed between the present and previous studies would be attributed to heterogeneity in the ecology and behavior of the vector.

Variation in EIR could also differ by mosquito collection methods [40]. [41] indicated that PSC might underestimate the HBR, which again underrates EIR. Previous studies also reported CDC light traps were more efficient than PSC to estimate EIR [21, 42–44]. Contrary to these, a study from Bioko Island, Equatorial Guinea, demonstrated that CDC light traps failed to determine the human-biting rate of the anthropogenic *A. gambiae* s.s [45]. Different from all previous reports, the present study indicated that higher EIRs were recorded from PSC catches than CDC light trap catches. Both CDC and PSC are reported to have shortcomings in mosquito sampling. While CDC light traps attract fed indoor-resting mosquitoes [3, 46], PSC tends to miss mosquitoes that leave the house after feeding including those entering the house after feeding outdoor [47]. Therefore, estimating the HBR using either CDC light trap or PSC has limitations, and the need to develop standard HBR remains high. Thus, the differences observed between the present and previous studies might be associated with limitation stated for each sampling method.

5. Conclusion

This study was linked with IRS application to assess its effect on EIR and other entomological risk factors for malaria transmission. The results illustrated that IRS was strong enough to reduce mosquito abundance, sporozoite rate, and EIR in areas having similar ecological setup with the present study villages [48].

Author details

Alemayehu Abate[1] and Melaku Wale[2*]

1 Ethiopian Public Health Institute, Addis Ababa University, Ethiopia

2 Department of Biology, Bahir Dar University, Ethiopia

*Address all correspondence to: melakuwale@gmail.com

IntechOpen

© 2019 The Author(s). Licensee IntechOpen. This chapter is distributed under the terms of the Creative Commons Attribution License (http://creativecommons.org/licenses/by/3.0), which permits unrestricted use, distribution, and reproduction in any medium, provided the original work is properly cited. cc BY

References

[1] Bruce-Chwatt LJ et al. Chemotherapy of Malaria. Revised. 2nd ed. Geneva: World Health Organization; 1986

[2] MacDonald G. The Epidemiology and Control of Malaria. London: Oxford University Press; 1957

[3] Lines JD, Curtis CF, Wilkes TJ, Njunwa KJ. Monitoring human-biting mosquitoes (Diptera: Culicidae) in Tanzania with light-traps hung beside mosquito nets. Bulletin of Entomological Research. 1991;**81**:77-84

[4] World Health Organization. Manual on Practical Entomology in Malaria. Part 2. Methods and Techniques. Geneva: World Health Organization; 1975

[5] Abose T, Yeebiyo Y, Olana D, Alamirew D, Beyene Y, Regassa L, et al. Re-orientation and Definition of the Role of Malaria Vector Control in Ethiopia. Geneva: World Health Organization; 1998

[6] Taye A, Hadis M, Adugna N, Tilahun D, Wirtz RA. Biting behaviour and Plasmodium infection rates of *Anopheles arabiensis* from Sille, Ethiopia. Acta Tropica. 2006;**97**:50-54

[7] FDREOH. National Malaria Guidelines. 3rd ed. Addis Ababa, Ethiopia: The Federal Democratic Republic of Ethiopia, Ministry of Health; 2012

[8] Cullen JR, De Zulueta J. Observations on the effect of residual insecticides in experimental huts in Masaka District, Uganda. Bulletin of the World Health Organization. 1964;**30**:263-278

[9] Abate A, Hadis H. Susceptibility of *Anopheles gambiae* s.l to DDT, malathion, permethrin and deltamethrin in Ethiopia. Tropical Medicine and International Health. 2011;**16**:486-491

[10] Beier JC, Killeen GF, Githure JI. Short report: Entomologic inoculation rates and *Plasmodium falciparum* malaria prevalence in Africa. The American Journal of Tropical Medicine and Hygiene. 1999;**61**(1):109-113

[11] WHO. World Malaria Report. 2011. Available from: http//www.who.int/entity/malaria/world_malaria_report_2011/9789244564403_eng.pdf

[12] World Health Organization. Malaria Entomology and Vector Control. Geneva: World Health Organization; 2003

[13] Garrett-Jones C. Prognosis for interruption of malaria transmission through the assessment of mosquito vectorial capacity. Nature. 1964a;**204**:1173

[14] Garrett-Jones C. The human blood index of malaria vectors in relation to epidemiological assessment. Bulletin of the World Health Organization. 1964b;**30**:241-261

[15] Drakeley C, Schellenberg D, Kihondaj J, Sousa CA, Arez AP, Lopes D, et al. An estimation of the entomological inoculation rate for Ifakara: A semi-urban area in a region of intense malaria transmission. Tropical Medicine & International Health. 2003;**8**:767-774

[16] Wirtz RA, Zavala F, Charoenvit Y, Cambell GH, Burkot TR, Shneider I, et al. Comparative testing of *Plasmodium falciparum* sporozoite monoclonal antibodies for ELISA development. Bulletin of the World Health Organization. 1987;**65**:39-45

[17] Chanda E, Phiri FN, Chanda J, Ramdeen VR, Kamuliwo M, Baboo KS. Impact of entomological interventions on malaria vector bionomics in low transmission settings in Zambia. Journal of Public Health

and Epidemiology. 2012;**4**(7):89-196. DOI: 10.5897/JPHE12.038. Available from: http://www.academicjournals.org/JPHE

[18] Alegana VA, Kigozi SP, Nankabiraw J, et al. Spatio-temporal analysis of malaria vector density from baseline through intervention in a high transmission setting. Parasites & Vectors. 2016;**9**:637. DOI: 10.1186/s13071-016-1917-3

[19] Habtewold T, Walker AR, Curtis CF, Osir EO, Thapa N. The feeding behaviour and *Plasmodium* infection of Anopheles mosquitoes in southern Ethiopia in relation to use of insecticide-treated livestock for malaria control. Transactions of the Royal Society of Tropical Medicine and Hygiene. 2001;**95**:584-586

[20] Hadis M, Lulu M, Makonnen Y, Asfaw T. Host choice by indoor resting *Anopheles arabiensis* in Ethiopia. Transactions of the Royal Society of Tropical Medicine and Hygiene. 1997;**91**:376-378

[21] Massebo F, Balkew M, Gebre-MichaelT,LindtjørnB.Entomologic inoculation rates of *Anopheles arabiensis* in Southwestern Ethiopia. The American Journal of Tropical Medicine and Hygiene. 2013;**89**(3):466-473. DOI: 10.4269/ajtmh.12-0745

[22] Waka M, Hopkins RJ, Akinpelu O, Curtis C. Transmission of malaria in the Tesseney area of Eritrea: Parasite prevalence in children, and vector density, host preferences, and sporozoite rate. Journal of Vector Ecology. 2005;**30**:27-31

[23] Collins FH, Mendez MA, RasmussenMO,MehaffeyPC,BesanskyNJ, Finnerty V. A ribosomal RNA gene probe differentiates member species of the *Anopheles gambiae* complex. The American Journal of Tropical Medicine and Hygiene. 1987;**37**:37-41

[24] Fornadel CM, Norris DE. Increased endophily by the malaria vector *Anopheles arabiensis* in southern Zambia and identification of digested blood meals. The American Journal of Tropical Medicine and Hygiene. 2008;**79**:876-880

[25] Mzilahowa T, Hastings JM, Molyne ME, McCall PJ. Entomological indices of malaria transmission in Chikhwawa district, Southern Malawi. Malaria Journal. 2012;**11**:380. http://www.malariajournal.com/content/11/1/380; Accessed in December 2017

[26] Thompson R, Begtrup K, Cuamba N, Dgedge M, Mendis C, Gamage-Mendis A, et al. The Matola malaria project: A temporal and spatial study of malaria transmission and disease in a suburban area of Maputo, Mozambique. American Journal of Tropical Medicine and Hygiene. 1997;**57**:550-559

[27] Coleman S, Dadzie SK, Seyoum A, et al. A reduction in malaria transmission intensity in Northern Ghana after 7 years of indoor residual spraying A reduction in malaria transmission intensity in Northern Ghana after 7 years of indoor residual spraying. Malaria Journal. 2017;**16**:324. DOI: 10.1186/s12936-017-1971-0

[28] Ossè RA, Aïkpon R, Gbédjissi GL, et al. A shift from indoor residual spraying (IRS) with bendiocarb to long-lasting insecticidal (mosquito) nets (LLINs) associated with changes in malaria transmission indicators in pyrethroid resistance areas in Benin. Parasites & Vectors. 2013;**6**(73). Available from: http://www.parasitesandvectors.com/content/6/1/73

[29] Animute A, Balkew M, Gebermichale T, Lindtjørn B. Blood meal sources and entomological inoculation rates of anophelines along a highland altitudinal transect in south-central Ethiopia. Malaria

Journal. 2013;**12**(76). Available from: http://www.malariajournal.com/content/12/1/76

[30] Taye B, Lelisa K, Emana D, Asale A, Yewhalaw D. Seasonal dynamics, longevity, and biting activity of Anopheline mosquitoes in Southwestern Ethiopia. Journal of Insect Science. 2016;**16**(1):1-7

[31] Beier JC, Perkins PV, Wirtz RA, Koros J, Dlggs D, Gargan TP, et al. Blood meal identification by direct enzyme-linked immunosorbent assay (ELISA), tested on *Anopheles* (Diptera: Culicidae) in Kenya. Journal of Medical Entomology. 1988;**25**:9-16

[32] Hay SI, Rogers DJ, Toomer JF, Snow RW. Annual *Plasmodium falciparum* entomological inoculation rates (EIR) across Africa: Literature survey, internet access and review. Transactions of the Royal Society of Tropical Medicine and Hygiene. 2000;**94**(2):113-127

[33] Ndenga B, Githeko A, Omukunda E, Munyekenye G, Atieli H, Wamai P, et al. Population dynamics of malaria vectors in Western Kenya highlands. Journal of Medical Entomology. 2006;**43**(2):200-206

[34] Mabaso MIH, Craig M, Ross A, Smith T. Environmental predictors of the seasonality of malaria transmission in Africa: The challenge. The American Journal of Tropical Medicine and Hygiene. 2007;**76**(1):33-38

[35] Shililu JI, Maier WA, Seitz HM, Orago AS. Seasonal density, sporozoite rates, and entomological inoculation rates of *Anopheles gambiae* and *Anopheles funestus* in high-altitude sugarcane growing zone in Western Kenya. Tropical Medicine & International Health. 1998;**3**:706-710

[36] Breman JG. The ears of the hippopotamus: manifestations, determinants, and estimates of the malaria burden. The American Journal of Tropical Medicine and Hygiene. 2001;**64**(1, 2S):1-11

[37] Kent RJ, Thuma PE, Mharakurwa S, Norris DE. Seasonality, blood feeding behavior, and transmission of *Plasmodium falciparum* by *Anopheles arabiensis* after an extended drought in southern Zambia. The American Journal of Tropical Medicine and Hygiene. 2007;**76**:267-274

[38] Himeidan Y, Elzaki MM, Kuleka EJ, Ibrahim M, Elhasson IM. Pattern of malaria transmission along the Rahad River basin, eastern Sudan. Parasites & Vectors. 2011;**4**:109

[39] Shililu JI, Ghebremeskel T, Mengistu S, Fekadu H, Zerom M, Mbogo C, et al. High seasonal variations in entomologic inoculation rate in Eritrea, a semi-arid region of unstable malaria in Africa. American Journal of Tropical Medicine and Hygiene. 2003;**69**:607-613

[40] Shaukat A, Breman JG, McKenzie FE. Using entomological inoculation rate to assess the impact of vector control on malaria parasite transmission and elimination. Malaria Journal. 2010;**9**:122

[41] Krafsur E. The bionomics and relative prevalence of *Anopheles* species with respect to the transmission of Plasmodium to man in western Ethiopia. Journal of Medical Entomology. 1977;**25**:180-194

[42] Fornadel CM, Norris LC, Glass GE, Norris DE. Analysis of *Anopheles arabiensis* blood feeding behavior in southern Zambia during the two years after introduction of insecticide-treated bed nets. The American Journal of Tropical Medicine and Hygiene. 2010;**83**:848-853

[43] Fornadel C, Norris LC, Norris DE. Centers for disease control light traps

for monitoring *Anopheles arabiensis* human biting rates in an area with low vector density and high insecticide treated bed net use. The American Journal of Tropical Medicine and Hygiene. 2010;**83**:838-842

[44] Ndiath M, Mazenot C, Gaye A, Konate L, Bouganali C, Faye O, et al. Methods to collect Anopheles mosquitoes and evaluate malaria transmission: A comparative study in two villages in Senegal. Malaria Journal. 2011;**10**(270):28

[45] Overgaard H, Sæbø S, Reddy MR, Reddy VP, Abaga S, Matias A, et al. Light traps fail to estimate reliable malaria mosquito biting rates on Bioko Island, Equatorial Guinea. Malaria Journal. 2012;**11**:56

[46] Petrarca V, Beier JC, Onyango F, Koros J, Asiago C, Koech DK, et al. Species composition of the *Anopheles gambiae* complex (Diptera: Culicidae) at two sites in western Kenya. Journal of Medical Entomology. 1991;**28**:307-313

[47] Mboera LE. Sampling techniques for adult Afrotropical malaria vectors and their reliability in the estimation of entomological inoculation rate. Tanzania Health Research Bulletin. 2005;**7**:117-124

[48] Massebo F, Balkew M, Gebermichael T, Lindtjørn B. Zoophagic behaviour of anopheline mosquitoes in southwest Ethiopia: Opportunity for malaria vector control. Parasites & Vectors. 2015;**8**:845. DOI: 10.1186/s13071-015-1264-9

Chapter 6

The Effects of Infection on Mosquito Rhythmic Behavior

Rafaela Vieira Bruno, Luana Cristina Farnesi and Luciana Ordunha Araripe

Abstract

Most organisms live in a rhythmic world, where daily environmental variation has a profound effect on their behavior and physiology. In addition to abiotic influence, interactions with other organisms that have their own particular cycles are also part of circadian rhythm formation. In this chapter, we present aspects of the biology of mosquito vectors, more precisely *Aedes aegypti*, which is a vector of arboviruses of great epidemiological importance, like dengue, Zika, and chikungunya. The successful transmission of the virus depends on the coordination of several behavioral and physiological traits involved in the virus-vector-host interaction. Thus, understanding the mechanisms of endogenous control of rhythmic traits of the mosquito vector and the impact that both environmental variation and virus infection can have on this regulation is key for a reliable estimate of the vectorial capacity. We discuss the infection-driven changes in traits used to calculate parameters of the vectorial capacity, and finally, we review the current knowledge on the molecular mechanisms underlying vector rhythmic behavior and the potential cellular targets of arbovirus infection.

Keywords: *Aedes aegypti*, arbovirus, behavior, vectorial capacity, physiology, neurotropism, Zika, dengue, chikungunya, circadian clocks

1. Introduction

1.1 The *Aedes aegypti* mosquito as a vector of arboviruses

Aedes aegypti (Diptera: Culicidae) is an insect belonging to the family Culicidae and subgenus *Stegomyia*. The species was originally found in Egypt, hence its specific name; but it is currently distributed worldwide, occurring mainly in tropical and subtropical regions [1, 2]. *Aedes aegypti* life cycle is composed of four phases: egg, four larval instars, pupa, and adult (**Figure 1**).

Since they spend most of their life cycle in water, mosquitoes are considered to be primarily aquatic; they gain the terrestrial environment only in adulthood, when they fly in order to seek for food and mates [3–5]. Easy to distinguish for taxonomists, *A. aegypti* is a dark-bodied mosquito in the adult phase, with white spots on the dorsal abdomen and legs and a white pattern composed of a lira-shaped drawing on its scutum (**Figure 2**) [1, 6].

The mosquito *A. aegypti* is considered a major disease vector in urban habitats, being able to host and transmit various arbovirus. Females of anautogenous

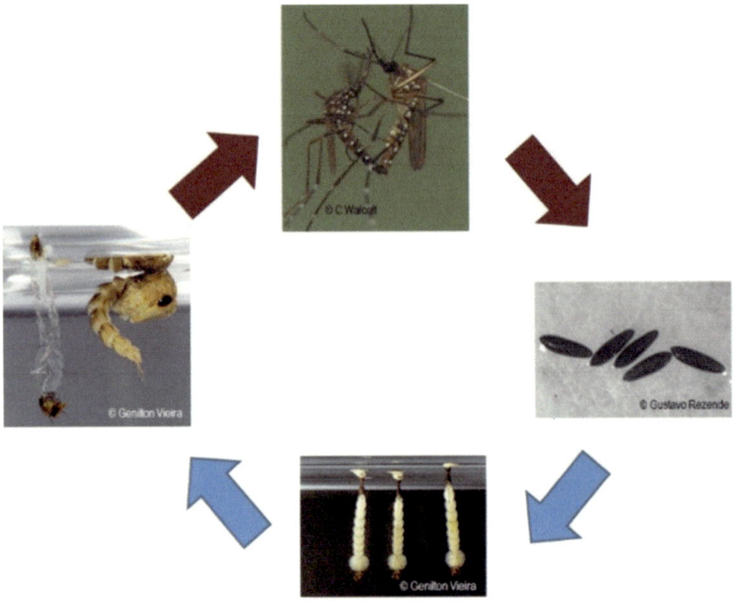

Figure 1.
Life cycle of the Aedes aegypti mosquito (credits of the photos on each image—out of ratio).

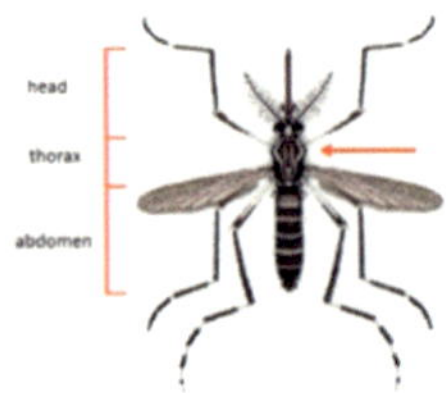

Figure 2.
Aedes aegypti mosquito. The arrow points to the scutum region that displays the lira-shaped (out of ratio). The three body parts, common to all insects, are indicated above. Adapted from [6].

mosquitoes, like *A. aegypti*, need a blood meal in order to mature their eggs and to perpetuate the species. Because of this, this mosquito is indubitably one of the most pathologically important arthropod vectors. One *A. aegypti* female is able to produce approximately 100 eggs after blood feeding on a vertebrate host, in each gonotrophic cycle (interval between blood meal and egg laying). Even being a diurnal mosquito, the female prefers to lay eggs in dark/shaded sites and in the dark phase of the day [7–10]. *Aedes aegypti* is an anthropophilic species, preferring to feed on human blood than the blood of other vertebrates [11, 12]. This feature is responsible for the role of *A. aegypti* as transmitter of many pathogens that cause important human diseases. The markedly anthropophilic and endophilic behaviors of *A. aegypti* make it a very efficient vector of yellow fever, dengue, chikungunya, and Zika viruses [3, 11, 13–16]. Many other factors related to behavior and physiology of vector and pathogens are significant for the success in arbovirus transmission, such as (i) the habit of laying eggs in multiple breeding sites; (ii) the diversity of posture sites; (iii) the gonotrophic discordance, that is, the ability to blood feed on more than one host for each batch of eggs produced; and (iv) the ability of the eggs in remaining viable, in quiescence state, for up to 1 year in dry conditions (called "egg resistance to desiccation") and in large temperature variation (i.e., 16–35°C) [1, 3, 17–20]. All these features can be associated to vector competence and vectorial capacity [7].

Both vector competence and vectorial capacity are critical for arbovirus transmission. Vector competence is the intrinsic ability of a vector to acquire, maintain, and transmit a pathogen to another host. In mosquitoes, a species is considered vector competent when females transmit the pathogen from one vertebrate to another during blood feeding [21]. This competence is related to intrinsic features of the vector, as well as the pathogen, such as pathogen genotype, pathogen strain, and vector strain. Specifically, for the viruses DENV and CHIKV, vector competence has been tested and confirmed in *A. aegypti* and *A. albopictus* laboratory strains. To DENV, environmental factors as daily temperature fluctuations have been demonstrated to impact vector-pathogen interactions, being able to modulate the *A. aegypti* competence to DENV transmission [15, 22, 23].

Although the number of studies on *A. aegypti* behavior and physiology, as well as arbovirus-mosquito interactions, has been growing, additional information is needed in order to promote the development of better mosquito control actions. Variation in the vector competence for different arboviruses highlights the existence of different virus-vector interactions. For example, both *A. aegypti* and *A. albopictus* show the competence to transmit the arboviruses DENV, ZIKV, YFV, and CHIKV; however these vectors do not exhibit the same transmission efficiency. Likewise, within-population genetic variation may explain the varied vector competence for different arboviruses and may also be related to the response of mosquitoes to control programs [24].

Other factors may be involved in the vector competence, for example, two different insecticide resistance mechanisms were described to enhance the vector competence of *Culex quinquefasciatus* for West Nile virus, which can impact on transmission dynamics of arboviruses for other mosquito vector species [25]. Measuring the vector competence of field mosquitoes for different arbovirus can help to assess the risk of arbovirus emergence [24].

Vectorial capacity, in turn, is the estimated value through a formula that takes into account a set of parameters of intraspecific physiology and behavior that, associated with environmental conditions, favor natural transmission of a given disease. The vectorial capacity is mainly influenced by population density, biting behavior (frequency of host contact for blood feeding), and mosquito vector survivorship [26]. The concept of vectorial capacity was initially established for the transmission of malaria by vectors of the genus *Anopheles* and calculated by the formula shown in **Figure 3**, where the total number of potentially infectious bites a day is one of the parameters. Many studies of mathematical models describing pathogen transmission by mosquitoes make similar assumptions [28, 29].

The World Health Organization emphasizes that mosquito vector control plays an important role in blocking the propagation of critical arboviruses. This is particularly relevant when no vaccines or specific drug treatments are available, as is the case for dengue, Zika, and chikungunya, which have the *A. aegypti* as the main vector [13, 14, 30]. Understanding vector competence and vector capacity mechanisms is important in designing safer vaccines and new strategies to prevent the transmission of pathogens. Specifically for the mosquito vector *A. aegypti*, many barriers hamper infection, dissemination, and transmission of arboviruses through mosquito vector tissues.

$$VC = ma^2\, bp^n / -\log_e p$$

Figure 3.
Vectorial capacity formula: here m is the number of female mosquitoes per host, a is the daily blood feeding rate, b is the transmission rate among exposed mosquitoes, p is the probability of daily survival, and n is the extrinsic incubation period (EIP). Adapted from Refs. [26, 27].

2. Aspects of mosquito behavior and their role on the vectorial capacity

In mosquitoes, locomotor activity [31–34], host-seeking and blood feeding [35], digestion, mate finding and reproduction, and site choice for oviposition [36–38] are examples of rhythmic patterns that are recognizably modulated by extrinsic factors [39]. While these patterns have been increasingly studied, ecological interactions between hematophagous females and their hosts and pathogens are not well understood [40]. Likewise, how female rhythms affect and are affected by males' biological aspects associated with courtship and mating is still obscure [41, 42]. An emerging field of study, namely, "the causes and consequences of daily rhythms in the interactions between vectors, their hosts and the pathogens they transmit," was reviewed in Rund et al. [40].

Cycles in behavior and physiology have coevolved so that the organism's fitness is optimized. A shift in the rhythm of these traits may disrupt important biological functions leading to impacts on fertility and viability. For instance, in *Drosophila*, a shift in the time of day that food is ingested leads to a reduction in fertility [43], whereas maintaining the expected time for food intake leads to the benefits of an improved cardiac function [44]. In mosquitoes, when females engage in foraging or seeking for hosts, a suite of enzymes responsible for blood digestion must be operating, their immune response must be on to avoid pathogen infection, and their detoxification against insecticides must be active [40]. Therefore, breaking the interlocked pathways for pathogen-vector-host interactions will affect the vectorial capacity and the epidemiology of arboviruses.

The vectorial capacity measures the chance of emergence of new cases of the disease departing from one infected human host. As such, the parameters of behavior and physiology used in the calculation assume that mosquitoes are infected. Other parameters include population density, frequency of bites [26, 40], and transmission competency, which are directly influenced by the vector's behavior and physiology, as well as by the pathogen's behavior and extrinsic incubation period (EIP) [26, 40, 45].

The magnitudes of most parameters of the vectorial capacity equation are highly dependent on the daily variation of locomotor/flight activity behavior. There are several ways of measuring the pattern of locomotor activity of insect species, varying from the traditional method of reporting the presence of one species in field traps, in different times of the day, to activity monitors and video imaging used in the laboratory. Data generated by all these methods are represented with similar graphics, where the amount of locomotor/flight activity registered at each time interval is plotted on a 24-h graph. Variation in activity is studied according to variation in a *Zeitgeber*, a term used for an environmental synchronizer such as light or temperature.

Field and laboratory studies show that *A. aegypti* is active during the day, with activity peaks at dawn and dusk and lack of activity at night [7, 9]. Because flight activity toward hosts is driven by olfactory signals in mosquitoes, one could expect that rhythms in the expression of odorant binding proteins should parallel the olfactory sensitivity to host odors in order to activate the behavioral output [40]. However, in *Aedes aegypti*, rhythms in olfactory sensitivity are not sufficient to explain the daily cycling in behavior toward hosts [39]. The authors performed electroantennography assays and Y-maze olfactometer experiments using five different volatiles (including plants and host odors) and found that the peak of olfactory behavior is decoupled from the variation in olfactory sensitivity. These results suggest that modulation of the behavior associated with olfactory cues happens in both the peripheral (antenna) and central (endogenous clock) levels.

Humans are the main hosts for *A. aegypti* females, and humans are most likely awake and active when these females are trying to land and blood feed. This imposes a risk for the mosquito. Body heat and carbon dioxide are the human factors that are the most attractive for mosquitoes [46]. The availability of these factors varies in a circadian way [40, 47, 48] and are subjected to changes in environmental conditions [40].

Light and temperature are the major environmental factors affecting the rhythmic behavior of most organisms. As such, variation in these factors has a profound effect on the vectorial capacity. For instance, the biting rate of *A. aegypti*, which is a fundamental parameter in the calculation of the vectorial capacity, is highly influenced by temperature and time of day. Since females need to be active in order to engage in blood seeking, temperatures below 15°C and above 36°C constrain locomotor activity and make the number and intensity of bites to cease [3, 49].

Mating interaction is another element influencing vectorial capacity. Significant alterations in females' physiology and behavior happen after copulation, when male accessory gland peptides are transferred along with sperm [50], though contrasting effects have been reported. Augmented host-seeking and blood-feeding activity [31, 51–53], as well as an increase in oviposition rates [54, 55] and egg development [56], were reported, suggesting that these alterations could boost up the vectorial capacity. On the other hand, Lima-Camara et al. [9] have found a significant decrease in the mean locomotor activity after insemination and after blood feeding in females of *A. aegypti*. Although this result was reported as the daily mean of locomotor activity, the occurrence of a significant increase in the dusk peak of activity, which is the peak associated with biological functions like host-seeking and oviposition, is remarkable.

3. The effects of infection on behavior and physiology of mosquito vectors

Since vectorial capacity suffers major influence of vector behavior, studying the degree of modulation that arbovirus exerts on *A. aegypti*'s behavior is a key factor for understanding infection dynamics and host pathogenesis. In a recent work, Gaburro et al. have shown that infection by Zika virus leads to neuro-excitation in *A. aegypti*'s brain, inducing changes in the mosquito's behavior. The increase in neuronal spikes in infected versus non-infected females reflected on an increase in flight activity when females were studied in pools [57]. The authors found replicating virus in ZIKV-infected female brains, characterizing the tropism for the central nervous system, as well as in sensory organs like antennas and eyes, potentially affecting neuronal communication. Likewise, dengue virus was also found to be neurotropic in mosquitoes [58].

A consequence of the neurotropic characteristic of these arboviruses is the alteration in the patterns of locomotor activity and feeding behavior. For instance, *A. triseriatus* becomes more avid for refeeding when infected by La Crosse virus [59, 60], while *Aedes aegypti* becomes more active when infected with serotype 2 of dengue virus [61] and with Zika virus when females are monitored in groups in cages (**Figure 4**) [57]. However, the assumption that virus infection would modulate behavior in a way to increase virus transmission and vectorial capacity is not always met. The example of West Nile virus indicates a possible decrease in virus transmission, where the mosquito vector *Culex pipiens* becomes less avid for host-seeking when infected with the virus [63]. Likewise, for individually monitored females of *A. aegypti*, Zika virus infection reduces flight activity, suggesting that infected mosquitoes may remain associated with closely distributed human hosts (**Figure 4**) [62].

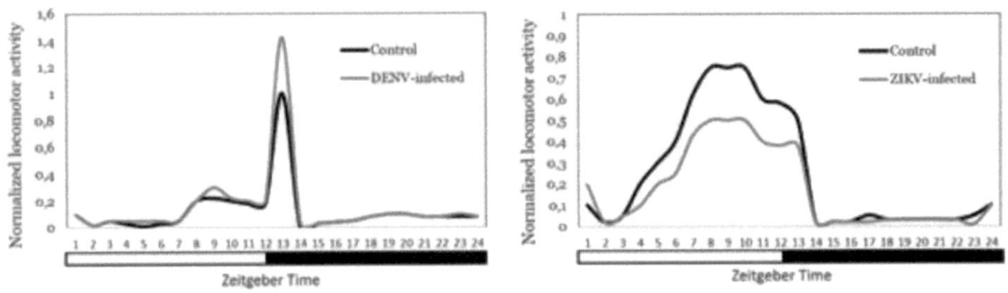

Figure 4.
Locomotor activity of virus-infected and not infected females. (A) Females infected with serotype 2 of dengue virus (modified from [61]). (B) Females infected with Zika virus (modified from [62]). The Zeitgeber time means the time passed, in hours, after light is turned on.

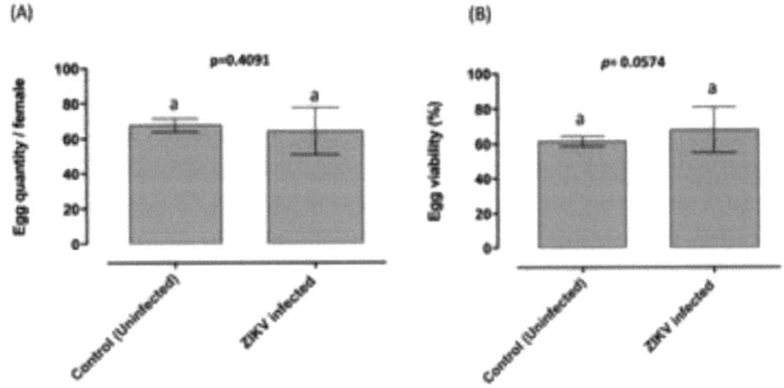

Figure 5.
Effect of Zika virus infection on the fecundity (A) and fertility (B) of Aedes aegypti females, on the third gonotrophic cycles. The lack of significance is represented by p values >0.05 obtained by using the nonparametric Mann-Whitney tests. Error bars represent mean ± s.d of three independent experiments (modified from [62]).

Arbovirus infection is also responsible for changes in physiological traits implicated in the estimate of the vectorial capacity. The number of female mosquitoes per host is one of the most important parameters of the vectorial capacity and is directly influenced by life history traits like the number of eggs laid by females (fecundity) and the number of viable offspring (fertility). These traits have been reported altered by arbovirus infection, although the effect varies depending on the virus. Dengue-infected females of *A. aegypti* produce a significantly lower number of eggs with a lower hatching rate [45], while ZIKV-infected and non-infected females do not show significant differences in fecundity and fertility (**Figure 5**) [62].

4. Human environmental impact and the effects on vector-host interaction and the risk of disease transmission

Human occupation may lead to profound alterations in the environment, such as global warming and light pollution. Some of these changes impose new selective pressures to all organisms involved with the infection, say pathogens, vectors, and hosts, but also their predators and the vegetation used as nutrition or habitat. The modeling of the effects of global warming on disease transmission indicates a shift in the global distribution of *Aedes*-borne virus with mild to severe effects on the risk of transmission [64, 65]. Concerning light pollution, the increase in domiciliary and peridomiciliary lighting may extend the phase of activity of *A. aegypti* by a couple of hours, which may raise the biting rate and the chance of arbovirus transmission. A recent work showed that artificial lighting at night make house sparrows, the reservoirs of West Nile virus, to become infectious for a period of 2 days longer than house sparrows that get dark nights [66]. This leads to an increase of 41% in the potential of disease outbreak. Comparatively, in light-night areas, nocturnal mosquitoes like *Anopheles* species will begin their phase of activity when human hosts are still active and not under a bed net, leading to a higher chance of malaria transmission [46].

Altogether, both vectors and hosts undergo behavioral and physiological changes triggered by the virus infection, and in turn, the influence of environmental variation is behind all facets of this interaction. The next section will discuss the endogenous mechanisms regulating rhythmic behavior and physiology, as well as the role of environmental factors on synchronizing these rhythms.

5. Molecular control of the behavior

The different behaviors exhibited by mosquitoes are, in general, driven by internal biological clocks that generate circadian rhythms. These rhythms present a period of nearly, but not exactly, 24 h and are responsible for responses such as host-seeking, breeding site seeking, activity, and rest, among others [67].

These rhythms are directly influenced by natural cues from the environment, and the most important ones are the light/dark and the temperature cycles. These stimuli are received by specific receptors, like photoreceptors (in the eyes and head) and thermoreceptors (along the whole body) and are transmitted to the internal pacemaker or the biological clock itself. Thus, a rhythm or a physiological response is generated from the interaction of the stimuli with the pacemaker neurons [68].

The pacemaker neurons are so-called because they express the clock genes, which are the components of the circadian clock. These genes interact with each other and recruit kinases, phosphatases, and transcription factors to generate oscillating expression in a 24-h cycle [69]. They are also responsible for the regulation of many other genes, the clock-controlled genes (CCGs), that are directly associated with tissue-specific functions [70].

Drosophila melanogaster is the insect model for studying circadian rhythm, but it is already known that many features of the circadian clock of other insects differ considerably from the fly clock. The *Drosophila* clock is formed by three interconnected autoregulatory loops, in which the proteins coded by *Clock* (*Clk*) and *cycle* (*cyc*) genes play a central role. In the first loop, the heterodimer CLK-CYC binds to an E-box sequence in the promoter region of *period* (*per*) and *timeless* (*tim*) genes, activating their transcription. Once in the cytoplasm, the transcripts are translated into proteins that accumulate during the early night and later enter the nucleus to repress their own transcription. This cycle lasts 24 h due to the posttranslational modifications controlled by the activity of kinases such as DOUBLETIME (DBT), CASEIN KINASE 2A (CK2A), and SHAGGY (SGG), which together with phosphatases such as PP2A stabilize PER and TIM [71, 72]. In the second loop, two transcription factors, VRI and PDP1e, are involved, respectively, in the repression and activation of *Clk* and *cyc* genes. Finally, a third interconnected loop involves the activation of *clockwork orange* (*cwo*) gene and the repression exerted by its product, CWO, in PER targets [73]. **Figure 6** summarizes the *D. melanogaster* molecular circadian clock.

An interesting feature of this clock is its property of environmental synchronization, which adjusts the period to exactly 24 h. One of the most important synchronizers (or *Zeitgebers*) is light. The light-induced resetting mechanism is dependent upon CRYPTOCHROME (CRY), which is a photoreceptor that induces TIM phosphorylation and leads it to degradation via proteossome [74, 75]. Other stimuli act as *Zeitgebers*, such as temperature and food, and their importance varies from species to species. In *A. aegypti*, temperature cycles are a very strong environmental cue, although the molecular mechanisms for entrainment are still unknown [76].

Molecular studies regarding the circadian clock in *A. aegypti* have been purely descriptive, because of the lack of genetic tools to unravel the function of clock genes in this species. Moreover, the phylogenetic distance between *Aedes aegypti* and *Drosophila melanogaster* implies significant differences between the two species in several biological aspects, beyond the circadian expression pattern. One significant difference is the presence, in *A. aegypti*, of a second cryptochrome gene, called *cryptochrome 2* (*cry2*), which does not exist in *D. melanogaster* [77]. This orthologue in mammals is a transcriptional repressor [78], and some studies done in *Danaus plexippus* confirmed this function in monarch butterflies [79]. Therefore, it is reasonable that this gene also plays this role in mosquitoes as well.

In a general manner, the circadian expression pattern of the main clock genes in *A. aegypti* under light dark cycles (LD12:12, which means 12 h of light followed by 12 h of dark) and constant temperature presents some similarities to what is observed in the *D. melanogaster* model. Genes *per* and *tim* present a cyclic mRNA expression with a peak in ZT 17, in the middle of the dark phase, and *vri* mRNA peak expression occurs some hours earlier than *Pdp1* mRNA peak expression (ZT 11 × ZT 17, respectively). However, two striking differences in relation to the *Drosophila* clock can be seen. The first one is related to the expression of the genes that encode the transcriptional activators, *Clk* and *cyc.* In Drosophila, *Clk* is the

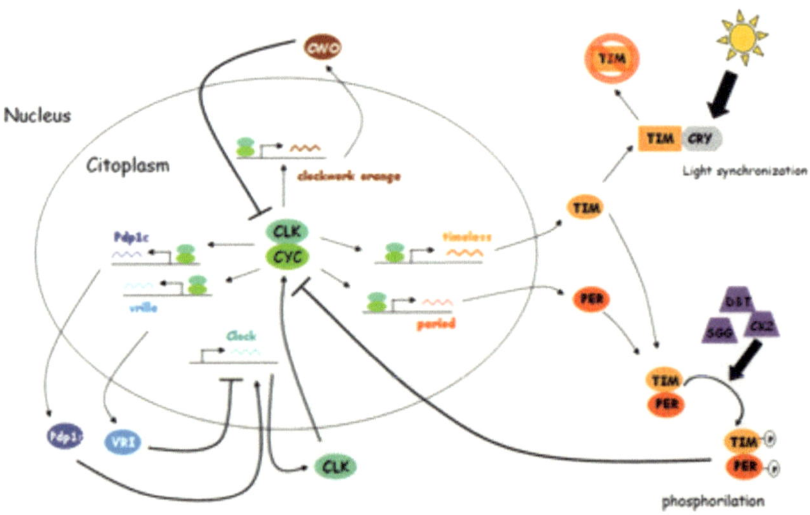

Figure 6.
The molecular circadian clock of Drosophila melanogaster. The heterodimer CLK-CYC plays a crucial role in the maintenance of the 24-h cycle, integrating the three regulatory feedback loops. In the first loop, CLK-CYC binds to the E-box regions in per and tim promoters and activates their transcription. Once in the cytoplasm, they are translated into proteins and suffer posttranslational modifications, caused by kinases and phosphatases. Once they can accumulate, the heterodimer PER-TIM enters in the nucleus and represses its own transcription. In the second loop, CLK-CYC heterodimer binds to E-boxes and activates vri *and* Pdp1ε *transcription. VRI binds to V/P boxes and inhibits* Clk *transcription, while PDP1ε accumulates a few hours later and shifts VRI from V/P box and activates Clk transcription. Finally, the third loop is related to the activation of* clockwork orange *gene and the inhibition of CLK-CYC by this gene product. The light synchronization is fired by a conformational change of CRY, which allows it to bind to TIM and leads to its degradation via proteasome, reinitiating the cycle. Straight arrows indicate following steps in the feedback loops; curved arrows indicate activation interactions; blocked line indicates inhibitory interaction. Colored forms represent proteins and wavy lines represent the mRNA. Based on [71, 72].*

main activator and presents a peak mRNA expression in antiphase to *per* and *tim*, whereas *cyc* presents a constitutive mRNA expression. On the contrary, *A. aegypti* Clk mRNA expression is arrhythmic, while *cyc* presents a very robust mRNA cycling pattern [8, 80]. Some years later, when *Clk* and *cyc* nucleotides and corresponding proteins were characterized, it was observed that the brain and muscle ARNT-like (BMAL) C-terminus region (BCTR) activation domain is missing in *Drosophila* CYC protein but is present in *A. aegypti* CYC protein [81]. These data can partially explain the variations in *Clk* and *cyc* expression pattern and suggest that there is a dissimilarity in circadian regulation between the two species. The second difference is related to the presence of the *cry2* gene and its bimodal mRNA expression pattern. As mentioned above, cry2 gene is present in *A. aegypti* and other *Diptera* but is absent in *D. melanogaster* [77]. According to putative clock models proposed by Yuan et al. in 2007, the presence or absence of *cryptochrome* genes may lead to crucial modifications in clock regulation.

5.1 The pacemaker neurons in insects

Clock genes are expressed in specific groups of neurons called pacemaker neurons, in the central nervous system of the organism, and are identified as pacemakers due to PER expression [82]. However, the distribution of these cells in the brain can vary from species to species; while in *Manduca sexta*, the pacemaker is located in the dorsal protocerebrum, in *Pachnoda marginata* it is located in the proximal optic lobe [82]. In *Drosophila*, around 150 clock neurons are located in the lateral protocerebrum, close to optic lobe and in the dorsal protocerebrum. The lateral neurons are subdivided in ventrolateral neurons (LNvs), dorsolateral neurons (LNds), lateral posterior neurons (LPNs), and dorsal neurons (DNs) [83]. Each group expresses the clock genes and communicates with each other according to different neurotransmitters, such as pigment-dispersing factor (PDF), which is the most well known [84]. Depending on the combination between neurotransmitters and hormone signaling among the neurons and clock genes' expression during the time of day, distinct neuronal clusters are responsible for different behaviors, such as feeding (which is regulated by the dorsolateral neurons) or temperature preference (regulated by dorsal neurons). On the other hand, the locomotor behavior, for instance, recruits all groups of neurons, which interact with each other to generate activity and rest along the 24-h day [72].

In *A. aegypti* there is not yet a map of the pacemaker neurons. However, it is described that some clustered neurons present a cycling expression of *cyc* and *per* mRNAs in antiphase, which strongly indicates that the areas where pacemaker neurons are found are similar to those observed in *D. melanogaster* [85]. Regarding the arbovirus infection, it was observed that both DENV and ZIKV are able to impact the neurons' spike activity in opposite manners. While ZIKV infection leads to hyperexcitation in a primary neuron culture, DENV2 infection does not seem to alter the spikes. Moreover, ZIKV reaches a plateau in replication around 2 dpi, whereas DENV2 initially decreases its replication and follows an increase in virus titers until 3 dpi [57]. It was already observed that ZIKV presents a strong neurotropism in mosquitoes [57], but it was not possible to associate the infected cells with clock neurons. Additional studies are necessary to establish the relationship between arbovirus infection and the pacemaker cells.

6. Conclusions

It is known that the virus-host interaction has a crucial importance in the spreading of a pathogen, since mutations in the viral genome or the genetic background of a mosquito population can enhance or even inhibit the replication of the virus in the mosquito. Beyond this genetic interaction, behavior is also directly related to the vectorial capacity of *A. aegypti*. Reports about the influence of the viral infection on several mosquitoes' behaviors have been increasing along the years. However, we still do not know the way arboviruses modulate the expression of the core clock genes that control behavior. It is even possible that infection does not have a direct effect on the core clock genes themselves but possibly on the genes regulated by them, leading to alterations in behavior and, consequently, impacts on the vectorial capacity. Improving the knowledge on behavioral traits that are susceptible to infection-driven changes (e.g., time-of-day biting activity, time-of-day mating behavior, time-of-day oviposition behavior) can increase the efficacy of strategies of vector control.

Author details

Rafaela Vieira Bruno[1,2]*, Luana Cristina Farnesi[1] and Luciana Ordunha Araripe[1]

1 Fundação Oswaldo Cruz-Fiocruz, Instituto Oswaldo Cruz, Laboratório de Biologia Molecular de Insetos, Rio de Janeiro, RJ, Brazil

2 Instituto Nacional de Ciência e Tecnologia em Entomologia Molecular, Rio de Janeiro, RJ, Brazil

*Address all correspondence to: rafaelav@ioc.fiocruz.br

IntechOpen

© 2019 The Author(s). Licensee IntechOpen. This chapter is distributed under the terms of the Creative Commons Attribution License (http://creativecommons.org/licenses/by/3.0), which permits unrestricted use, distribution, and reproduction in any medium, provided the original work is properly cited.

References

[1] Consoli R, Lourenço-de-Oliveira R. Principais Mosquitos de importância sanitária no Brasil. Rio de Janeiro: Fiocruz; 1994. 228 pp

[2] Bhatt S, Gething PW, Brady OJ, Messina JP, Farlow AW, et al. The global distribution and burden of dengue. Nature. 2013;**496**:504-507. DOI: 10.1038/nature12060

[3] Christophers S. *Aedes aegypti* (L.), the Yellow Fever Mosquito. Its Life History, Bionomics and Structure. Cambridge: The University Press; 1960. 739 pp

[4] Borror JD, Delong DM. Introdução ao estudo dos insetos. São Paulo: Edgar Blucher; 1988. 653 pp

[5] Forattini OP. Culicidologia Médica. Vol. 2. São Paulo: Editora da Universidade de São Paulo; 1996. 860 pp

[6] Rueda LM. Pictorial Keys for the Identification of Mosquitoes (Diptera: Culicidae) Associated with Dengue Virus Transmission. Zootaxa. Vol. 589. Auckland, New Zealand: Magnolia Press; 2004. ISBN 1-877354-46-5

[7] Clements AN. The Biology of Mosquitoes: Sensory Reception and Behaviour. London: Chapman and Hall; 1999. 740 pp

[8] Gentile C, Rivas GB, Meireles-Filho AC, Lima JB, Peixoto AA. Circadian expression of clock genes in two mosquito disease vectors: cry2 is different. Journal of Biological Rhythms. 2009;**24**(6):444-451. DOI: 10.1177/0748730409349169

[9] Lima-Camara TN, Lima JB, Bruno RV, Peixoto AA. Effects of insemination and blood-feeding on locomotor activity of *Aedes albopictus* and *Aedes aegypti* (Diptera: Culicidae) females under laboratory conditions. Parasites & Vectors. 2014;**7**(1):304. DOI: 10.1186/1756-3305-7-304

[10] Farnesi LC, Barbosa CS, Araripe LO, Bruno RV. The influence of a light and dark cycle on the egg laying activity of *Aedes aegypti* (Linnaeus, 1762) (Diptera: Culicidae). Memórias do Instituto Oswaldo Cruz. 2018;**113**:e170362. DOI: 10.1590/0074-02760170362

[11] Clements A. The Biology of Mosquitoes: Development, Nutrition and Reproduction. London: Chapman and Hall; 1992. 509 pp

[12] McMeniman CJ, O'Neill SL. A virulent Wolbachia infection decreases the viability of the dengue vector *Aedes aegypti* during periods of embryonic quiescence. PLoS Neglected Tropical Diseases. 2010;**4**:e748

[13] Boorman JP, Porterfield JS. A simple technique for infection of mosquitoes with viruses; transmission of Zika virus. Transactions of the Royal Society of Tropical Medicine and Hygiene. 1956;**50**(3):238-242

[14] Valle D, Pimenta DN, Cunha RV. Dengue: Teorias e Práticas. Rio de Janeiro: Fiocruz; 2015. 458 pp

[15] Costa-da-Silva AL, Ioshino RS, de Araújo HRC, Kojin BB, de Andrade Zanotto PM, Oliveira DBL, et al. Laboratory strains of *Aedes aegypti* are competent to brazilian Zika virus. PLoS One. 2017;**12**:e0171951

[16] Souza-Neto JA, Powell JR, Bonizzoni M. *Aedes aegypti* vector competence studies: A review. Infection, Genetics and Evolution. 2018;**67**:191-209

[17] Kliewer JW. Weight and hatchability of *Aedes aegypti* eggs (Diptera: Culicidae). Annals of the Entomological Society of America. 1961;**54**:912-917

[18] Rezende GL, Martins AJ, Gentile C, Farnesi LC, Pelajo-Machado M, et al. Embryonic desiccation resistance in *Aedes aegypti*: Presumptive role of the chitinized serosal cuticle. BMC Developmental Biology. 2008;**8**:82. DOI: 10.1186/1471-213X-8-82

[19] Farnesi LC, Martins AJ, Valle D, Rezende GL. Embryonic development of *Aedes aegypti* (Diptera:Culicidae): Influence of different constant temperatures. Memórias do Instituto Oswaldo Cruz. 2009;**104**:124-126. DOI: 10.1590/S0074-02762009000100020

[20] Farnesi LC, Vargas HCM, Valle D, Rezende GL. Darker eggs of mosquitoes resist more to dry conditions: Melanin enhances serosal cuticle contribution in egg resistance to desiccation in Aedes, Anopheles and Culex vectors. PLoS Neglected Tropical Diseases. 2017;**11**:e0006063. DOI: 10.1371/journal.pntd.0006063

[21] WHO. Technical Report Series No. 205. Malaria: Eighth Report of the Expert Committee. Geneva: WHO; 1961

[22] Lambrechts L, Paaijmans KP, Fansiri T, Carrington LB, Kramer LD, Thomas MB, et al. Impact of daily temperature fluctuations on dengue virus transmission by *Aedes aegypti*. Proceedings of the National Academy of Sciences of the United States of America. 2011;**108**(18):7460-7465. DOI: 10.1073/pnas.1101377108

[23] Chouin-Carneiro T, Vega-Rua A, Vazeille M, Yebakima A, Girod R, Goindin D, et al. Differential susceptibilities of *Aedes aegypti* and *Aedes albopictus* from the Americas to Zika virus. PLoS Neglected Tropical Diseases. 2016;**10**:e0004543

[24] Amraoui F, Ben Ayed W, Madec Y, Faraj C, Himmi O, Btissam A, et al. Potential of *Aedes albopictus* to cause the emergence of arboviruses in Morocco. PLoS Neglected Tropical Diseases. 2019;**13**:e0006997

[25] Atyame Célestine M, Haoues A, Laurence M, Vazeille M, Diallo M, Mylène W, et al. Insecticide resistance genes affect *Culex quinquefasciatus* vector competence for West Nile virus. Proceedings of the Royal Society B. 2019;**286**(1894):20182273. DOI: 10.1098/rspb.2018.2273

[26] Kramer LD, Ciota AT. Dissecting vectorial capacity for mosquito-borne viruses. Current Opinion in Virology. 2015;**15**:112-118

[27] Macdonald G. Epidemiologic models in studies of vector borne diseases. Public Health Reports. 1961;**76**:753-764

[28] Garrett-Jones C. The human blood index of malaria vectors in relation to epidemiological assessment. Bulletin of the World Health Organization. 1964;**30**:241-261

[29] Smith DL, Dushoff J, McKenzie FE. The risk of a mosquito-borne Infection in a heterogeneous environment. PLoS Biology. 2004;**2**(11):e368

[30] McSweegan E, Weaver SC, Lecuit M, Frieman M, Morrison TE, Hrynkow S. The global virus network: Challenging chikungunya. Antiviral Research. 2015;**120**:147-152. DOI: 10.1016/j.antiviral.2015.06.003

[31] Taylor B, Jones MD. The circadian rhythm of flight activity in the mosquito *Aedes aegypti* (L.). The phase-setting effects of light-on and light-off. The Journal of Experimental Biology. 1969;**51**(1):59-70

[32] Jones MDR, Hill M, Hope AM. The circadian flight activity of the mosquito, Anopheles gambiae: Phase setting by the light regime. The Journal of Experimental Biology. 1967;**47**:503-511

[33] Rowland M. Changes in the circadian flight activity of the mosquito Anopheles stephensi associated with insemination, blood-feeding, oviposition and nocturnal light intensity. Physiological Entomology. 1989;**14**:77-84. DOI: 10.1111/j.1365-3032.1989.tb00939.x

[34] Peterson EL. Phase-resetting a mosquito circadian oscillator. Journal of Comparative Physiology. 1980;**138**(3): 201-211. DOI: 10.1007/BF00657038

[35] Fritz ML, Walker ED, Yunker AJ, Dworkin I. Daily blood feeding rhythms of laboratory-reared north American *Culex pipiens*. Journal of Circadian Rhythms. 2014;**12**:1. DOI: 10.1186/1740-3391-12-1

[36] Sumba LA, Okoth K, Deng AL, Githure J, Knols BGJ, Beier JC, et al. Daily oviposition patterns of the African malaria mosquito *Anopheles gambiae* Giles (Diptera: Culicidae) on different types of aqueous substrates. Journal of Circadian Rhythms. 2004;**2**:6. DOI: 10.1186/1740-3391-2-6

[37] Fritz ML, Huang J, Walker ED, Bayoh MN, Vulule J, Miller JR. Ovipositional periodicity of caged *Anopheles gambiae* individuals. Journal of Circadian Rhythms. 2008;**6**:2. DOI: 10.1186/1740-3391-6-2

[38] Chadee DD, Ritchie SA. Oviposition behaviour and parity rates of *Aedes aegypti* collected in sticky traps in Trinidad, West Indies. Acta Tropica. 2010;**116**:212-216. DOI: 10.1016/j.actatropica.2010.08.008

[39] Eilerts DF, Giessen MV, Bose EA, Broxton K, Vinauger C. Odor-specific daily rhythms in the olfactory sensitivity and behavior of *Aedes aegypti* mosquitoes. Insects. 2018;**9**:147. DOI: 10.3390/insects9040147

[40] Rund SSC, O'Donnell AJ, Gentile JE, Reece SE. Daily rhythms in mosquitoes and their consequences for malaria transmission. Insects. 2016;7:14. DOI: 10.3390/insects7020014

[41] Alfonso-Parra C, Ahmed-Braimah YH, Degner EC, Avila FW, Villarreal SM, Pleiss JA, et al. Mating-induced transcriptome changes in the reproductive tract of female *Aedes aegypti*. PLoS Neglected Tropical Diseases. 2016;**10**(2):e0004451. DOI: 10.1371/journal.pntd.0004451

[42] Araripe LO, Bezerra JR, Rivas GBS, Bruno RV. Locomotor activity in males of *Aedes aegypti* can shift in response to females' presence. Parasites and Vectors. 2018;**11**(1):254. DOI: 10.1186/s13071-018-2635-9

[43] Xu K, Di Angelo JR, Hughes ME, Hogenesch JB, Sehgal A. The circadian clock interacts with metabolic physiology to influence reproductive fitness. Cell Metabolism. 2011;**13**:639-654. DOI: 10.1016/j.cmet.2011.05.001

[44] Gill S, Le HD, Melkani GC, Panda S. Time-restricted feeding attenuates age-related cardiac decline in Drosophila. Science. 2015;**347**:1265-1269. DOI: 10.1126/science.1256682

[45] Maciel-de-Freitas R, Koella JC, Lourenço-de-Oliveira R. Lower survival rate, longevity and fecundity of *Aedes aegypti* (Diptera: Culicidae) females orally challenged with dengue virus serotype 2. Transactions of the Royal Society of Tropical Medicine and Hygiene. 2011;**105**:452-458

[46] Knols BG, Meuerink J. Odors influence mosquito behavior. Scientia Medica. 1997;**4**:56-63

[47] Rund SSC, Bonar NA, Champion MM, Ghazi JP, Houk CM, et al. Daily rhythms in antennal protein and olfactory sensitivity in the malaria mosquito *Anopheles gambiae*. Scientific Reports. 2013;**3**:2494

[48] Bailey SL, Heitkemper MM. Circadian rhythmicity of cortisol and body temperature: Morningness-eveningness effects. Chronobiology International. 2001;**18**:249-261

[49] Aragao BH. Mosquitoes and yellow fever virus. Memórias do Instituto Oswaldo Cruz. 1939;**34**:565-581. DOI: 10.1590/S0074-02761939000400007

[50] Jones MDR. The programming of circadian flight activity in relation to mating and the gonotrophic cycle in the mosquito *Aedes aegypti*. Physiological Entomology. 1981;**87**:511-521. DOI: 10.1111/j.1365-3032.1981.tb00275.x

[51] Adlakha V, Pillai MKK. Role of male accessory gland substance in the regulation of blood intake by mosquitoes. Journal of Insect Physiology. 1976;**22**:1441-1442

[52] Lee J-J, Klowden MJ. A male accessory gland protein that modulates female mosquito (Diptera: Culicidae) host-seeking behavior. Journal of the American Mosquito Control Association. 1999;**15**:4-7

[53] Klowden MJ. The check is in the male: Male mosquitoes affect female physiology and behaviour. Journal of the American Mosquito Control Association. 1999;**15**:213-220

[54] Leahy MG, Craig GB. Accessory gland substance as a stimulant for oviposition in *Aedes aegypti* and *Aedes albopictus*. Mosquito News. 1965;**25**:448-452

[55] Hiss EA, Fuchs MS. The effect of matrone on oviposition in the mosquito, *Aedes aegypti*. Journal of Insect Physiology. 1972;**18**:2217-2227. DOI: 10.1016/0022-1910(72)90250-8

[56] Wallis RC, Lang CA. Egg formation and oviposition in bloodfed *Aedes aegypti*. Mosquito News. 1956;**16**:283-286

[57] Gaburro J, Bhatti A, Harper J, Jeanne I, Dearnley M, Green D, et al. Neurotropism and behavioral changes associated with Zika infection in the vector *Aedes aegypti*. Emerging Microbes and Infections. 2018;**7**(1):68. DOI: 10.1038/s41426-018-0069-2

[58] Salazar MI, Richardson JH, Sánchez-Vargas I, Olson KE, Beaty BJ. Dengue virus type 2: Replication and tropisms in orally infected *Aedes aegypti* mosquitoes. BMC Microbiology. 2007;**7**:9. DOI: 10.1186/1471-2180-7-9

[59] Jackson BT, Brewster CC, Paulson SL. La Crosse virus infection alters blood feeding behavior in Aedes triseriatus and Aedes albopictus (Diptera: Culicidae). Journal of Medical Entomology. 2012;**49**:1424-1429

[60] Spengler CM, Czeisler CA, Shea SA. An endogenous circadian rhythm of respiratory control in humans. The Journal of Physiology. 2000;**526**:683-694

[61] Lima-Camara TN, Bruno RV, Luz PM, Castro MG, Lourenço-de-Oliveira R, Sorgine MH, et al. Dengue infection increases the locomotor activity of *Aedes aegypti* females. PLoS One. 2011;**6**:e17690. DOI: 10.1371/journal.pone.0017690.t001

[62] Padilha KP, Resck ME, Cunha OA, Teles-de-Freitas R, Campos SS, Sorgine MH, et al. Zika infection decreases *Aedes aegypti* locomotor activity but does not influence egg production or viability. Memórias do Instituto Oswaldo Cruz. 2018;**113**(10):e180290

[63] Vogels CBF, Fros JJ, Pijlman GP, van Loon JJA, Gort G, Koenraadt CJM. Virus interferes with host-seeking behaviour of mosquito. The Journal of Experimental Biology. 2017;**220**(19):3598-3603. DOI: 10.1242/jeb.164186

[64] Carlson CJ, Dougherty ER, Getz W. An ecological assessment of the pandemic threat of Zika virus. PLoS Neglected Tropical Diseases. 2016;**10**:e0004968. DOI: 10.1371/journal.pntd.0004968

[65] Ryan SJ, Carlson CJ, Mordecai EA, Johnson LR. Global expansion and redistribution of Aedes-borne virus transmission risk with climate change. PLoS Neglected Tropical Diseases. 2019;**13**(3):e0007213. DOI: 10.1371/journal.pntd.0007213

[66] Kernbach ME, Newhouse DJ, Miller JM, Hall RJ, Gibbons J, Oberstaller J, et al. Light pollution increases West Nile virus competence of a ubiquitous passerine reservoir species. Proceedings of the Royal Society B. 2019;**286**:20191051. DOI: 10.1098/rspb.2019.1051

[67] Saunders DS. Insect Clocks. 3rd ed. Amsterdam: Elsevier; 2002. 560 pp

[68] Moore-Ede MC, Sulzman FM, Fuller CA. The Clocks that Time Us. Cambridge: Harvard University Press; 1982

[69] Mendoza-Viveros L, Bouchard-Cannon P, Hegazi S, Cheng AH, Pastore S, Cheng HM. Molecular modulators of the circadian clock: Lessons from flies and mice. Cellular and Molecular Life Sciences. 2017;**74**(6):1035-1059. DOI: 10.1007/s00018-016-2378-8

[70] Stanewsky R. Clock mechanisms in Drosophila. Cell and Tissue Research. 2002;**309**(1):11-26

[71] Hardin PE. Molecular genetic analysis of circadian timekeeping in Drosophila. Advances in Genetics. 2011;**74**:141-173. DOI: 10.1016/B978-0-12-387690-4.00005-2

[72] Franco DL, Frenkel L, Ceriani MF. The underlying genetics of Drosophila circadian behaviors. Physiology (Bethesda, Md.). 2018;**33**(1):50-62. DOI: 10.1152/physiol.00020.2017

[73] Kadener S, Stoleru D, McDonald M, Nawathean P, Rosbash M. Clockwork Orange is a transcriptional repressor and a new Drosophila circadian pacemaker component. Genes & Development. 2007;**21**(13):1675-1686

[74] Emery P, So WV, Kaneko M, Hall JC, Rosbash M. CRY, a Drosophila clock and light-regulated cryptochrome, is a major contributor to circadian rhythm resetting and photosensitivity. Cell. 1998;**95**:669-679

[75] Stanewsky R, Kaneko M, Emery P, Beretta B, Wager-Smith K, Kay SA, et al. The cryb mutation identifies cryptochrome as a circadian photoreceptor in Drosophila. Cell. 1998;**95**:681-692

[76] Rivas GBS, Teles-de-Freitas R, Pavan MG, Lima JBP, Peixoto AA, Bruno RV. Effects of light and temperature on daily activity and clock gene expression in two mosquito disease vectors. Journal of Biological Rhythms. 2018;**33**(3):272-288. DOI: 10.1177/0748730418772175

[77] Yuan Q, Metterville D, Briscoe AD, Reppert SM. Insect cryptochromes: Gene duplication and loss define diverse ways to construct insect circadian clocks. Molecular Biology and Evolution. 2007;**24**(4):948-955

[78] Nangle SN, Rosensweig C, Koike N, Tei H, Takahashi JS, Green CB, et al. Molecular assembly of the period-cryptochrome circadian transcriptional repressor complex. eLife. 2014;**3**:e03674. DOI: 10.7554/eLife.03674

[79] Zhu H, Sauman I, Yuan Q, Casselman A, Emery-Le M, Emery P, et al. Cryptochromes define a novel circadian clock mechanism in monarch butterflies that may underlie sun

compass navigation. PLoS Biology. 2008;**6**(1):e4. DOI: 10.1371/journal.pbio.0060004

[80] Ptitsyn AA, Reyes-Solis G, Saavedra-Rodriguez K, Betz J, Suchman EL, Carlson JO, et al. Rhythms and synchronization patterns in gene expression in the *Aedes aegypti* mosquito. BMC Genomics. 2011;**12**:153. DOI: 10.1186/1471-2164-12-153

[81] Chahad-Ehlers S, Arthur LP, Lima ALA, Gesto JSM, Torres FR, Peixoto AA, et al. Expanding the view of clock and cycle gene evolution in Diptera. Insect Molecular Biology. 2017;**26**(3): 317-331. DOI: 10.1111/imb.12296

[82] Helfrich-Förster C. Organization of endogenous clocks in insects. Biochemical Society Transactions. 2005;**33**(Pt 5):957-961

[83] Helfrich-Förster C, Yoshii T, Wülbeck C, Grieshaber E, Rieger D, Bachleitner W, et al. The lateral and dorsal neurons of *Drosophila melanogaster*: New insights about their morphology and function. Cold Spring Harbor Symposia on Quantitative Biology. 2007;**72**:517-525. DOI: 10.1101/sqb.2007.72.063

[84] Peng Y, Stoleru D, Levine JD, Hall JC, Rosbash M. Drosophila free-running rhythms require intercellular communication. PLoS Biology. 2003;**1**(1):E13. Epub 2003 Sep 15

[85] Chahad-Ehlers S, Gentile C, Lima JB, Peixoto AA, Bruno RV. Analysis of cycle gene expression in *Aedes aegypti* brains by in situ hybridization. PLoS One. 2013;**8**(1):e52559. DOI: 10.1371/journal.pone.0052559

Chapter 7

Metabarcoding: A Powerful Yet Still Underestimated Approach for the Comprehensive Study of Vector-Borne Pathogen Transmission Cycles and Their Dynamics

Anette Hernández-Andrade, Joel Moo-Millan, Nohemi Cigarroa-Toledo, Angel Ramos-Ligonio, Claudia Herrera, Bruno Bucheton, Jean-Mathieu Bart, Vincent Jamonneau, Anne-Laure Bañuls, Christophe Paupy, David Roiz, Denis Sereno, Carlos N. Ibarra-Cerdeña, Carlos Machaín-Williams, Julián García-Rejón, Sébastien Gourbière, Christian Barnabé, Jenny Telleria, Bruno Oury, Frédérique Brenière, Frédéric Simard, Miguel Rosado, Philippe Solano, Eric Dumonteil and Etienne Waleckx

Abstract

The implementation of sustainable control strategies aimed at disrupting the transmission of vector-borne pathogens requires a comprehensive knowledge of the vector ecology in the different eco-epidemiological contexts, as well as the local pathogen transmission cycles and their dynamics. However, even when focusing only on one specific vector-borne disease, achieving this knowledge is highly challenging, as the pathogen may exhibit a high genetic diversity and multiple vector species or subspecies and host species may be involved. In addition, the development of the pathogen and the vectorial capacity of the vectors may be affected by their midgut and/or salivary gland microbiome. The recent advent of Next-Generation Sequencing (NGS) technologies has brought powerful tools that can allow for the simultaneous identification of all these essential components, although their potential is only just starting to be realized. We present a metabarcoding approach that can facilitate the description of comprehensive host-pathogen networks, integrate important microbiome and coinfection data, identify at-risk situations, and disentangle the transmission cycles of vector-borne pathogens.

IntechOpen

This powerful approach should be generalized to unravel the transmission cycles of any pathogen and their dynamics, which in turn will help the design and implementation of sustainable, effective, and locally adapted control strategies.

Keywords: vector-borne diseases, transmission cycles, vector ecology, behavior, metabarcoding, next-generation sequencing (NGS), blood meals, microbiome, EcoHealth, One Health

1. Introduction

Vector-borne diseases affecting human health are caused by pathogens transmitted by "living organisms" between humans or from animals to humans. These "living organisms" are known as "vectors," which generally are bloodsucking arthropods, such as mosquitoes, ticks, flies, sandflies, fleas, or triatomine bugs. These arthropods ingest disease-producing microorganisms during a blood meal from an infected host (human or animal) and later transmit it to a new host during their subsequent blood meals [1]. According to the World Health Organization (WHO), vector-borne diseases, such as malaria, dengue, human African trypanosomiasis, leishmaniasis, Chagas disease, yellow fever, Japanese encephalitis, or onchocerciasis, account for almost 20% of all infectious diseases worldwide. They cause more than 700,000 deaths annually, and more than half of the world's population is estimated to be at risk of these diseases [1]. They are a major obstacle to development, and the poorest segments of societies and least-developed countries are the most affected. The most deadly vector-borne disease, malaria, causes more than 400,000 deaths annually, mainly children under 5 years. However, the world's fastest-growing vector-borne disease is dengue, with a 30-fold increase in disease incidence over the last 50 years [1, 2]. Currently, there is an estimation of 96 million cases of dengue per year, and more than 3.9 billion people in over 128 countries are at risk of contracting this disease [1, 3]. Chagas disease, which is one of the primary study models of our research group and classified by the WHO within the group of Neglected Tropical Diseases (NTDs), is a major public health problem in Latin America where 6–7 million people are currently infected [4, 5].

The control of vector-borne diseases relies mainly on control programs targeted against the different vectors. Nevertheless, the efficiency of the different vector control strategies is highly linked to the local ecology of the vectors [6], which in turn defines local transmission cycles. Consequently, for the implementation of sustainable control strategies aimed at disrupting the transmission of vector-borne pathogens, comprehensive knowledge of the vector ecology and behavior in the different eco-epidemiological contexts, as well as the local transmission cycles of the pathogens and their dynamics, is an essential need. However, even when focusing only on one specific vector-borne disease, achieving this knowledge is challenging. Indeed, the pathogen may exhibit a high genetic diversity, and multiple vector species or subspecies and host species may be involved. In addition, the development of the pathogen and the vectorial capacity of the vectors may be affected by their midgut and/or salivary gland microbiome. Sometimes, many pathogen species can also be involved. For example, leishmaniases are caused by more than 20 *Leishmania* species [7].

The recent advent of Next-Generation Sequencing (NGS) technologies has brought powerful tools, with enormous potential, allowing the simultaneous identification of all these components for the understanding of the eco-epidemiology of vector-borne diseases. Nevertheless, their potential is only just starting to be realized. Here, we present a metabarcoding approach based on NGS that can facilitate the creation of comprehensive host-pathogen networks, integrate important microbiome and coinfection data, identify at-risk situations, and disentangle the transmission cycles of vector-borne pathogens.

2. Complexity of vector-borne pathogen transmission cycles and their dynamics

The transmission cycles of vector-borne pathogens are shaped by the ecology and behavior of hosts and vectors in their specific environments and defined by the specific interactions between the vectors, the pathogens, and their hosts (which also act as blood-feeding sources of the vectors) [8]. Consequently, the comprehensive identification of these interactions is critical to disentangle transmission cycles and understand their dynamics. In most cases, an extraordinary diversity of organisms is involved, making the identification of those interactions challenging. In the case of Chagas disease, for example, the causative agent, a protozoan parasite called *Trypanosoma cruzi*, presents a very high genetic diversity, which has been classified into seven discrete typing units (DTUs) [9]. These DTUs are transmitted by more than 140 triatomine species, which live in a very wide variety of ecotopes and bioclimatic conditions [10], to more than 180 mammalian species, including wild animals, domestic animals, and human [11, 12]. In parallel, triatomines also take blood meals upon animals which are refractory to *T. cruzi* infection, called incompetent hosts, such as birds, reptiles, and amphibians [13, 14] (**Figure 1**). Finally, the establishment and development of the parasite and the vectorial capacity of the triatomines could be affected by the composition of their midgut microbiome [15], as has been shown for other vectors. For example, the development of *Trypanosoma brucei*, the agent of African trypanosomiasis, in its tsetse fly vector, is directly influenced by a microbiome-regulated gut immune barrier [16]. In the same way,

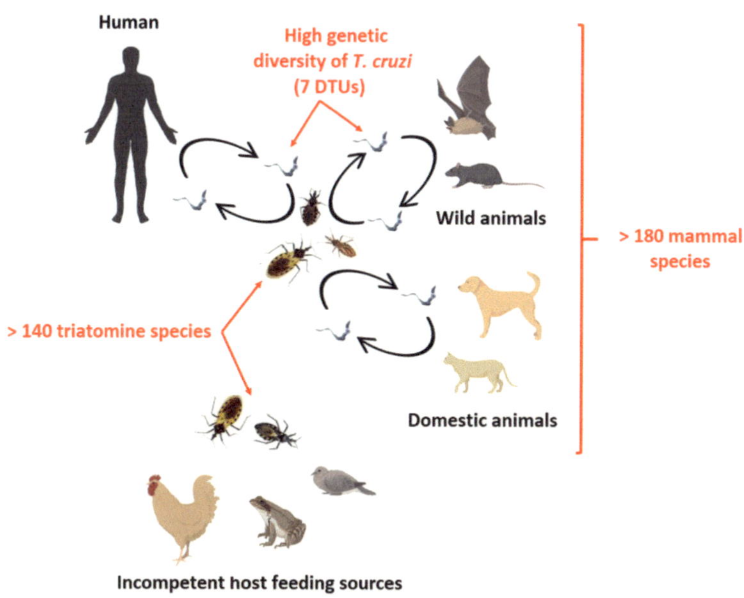

Figure 1.
Complexity of T. cruzi *transmission cycles. The parasite is divided into seven genetic subgroups (DTUs), which are transmitted by more than 140 triatomine species to more than 180 mammalian species, including wild animals, domestic animals, and human. In parallel, triatomines also take blood meals upon animals which are refractory to* T. cruzi *infection (incompetent hosts). Figure adapted from [25].*

the sand fly midgut microbiome is a critical factor for *Leishmania* growth and differentiation to its infective state prior to disease transmission [17]. Gut microbiome similarly modulates dengue virus infection in *Aedes aegypti* mosquitoes [18, 19], and microbiome manipulation may be used to control virus transmission [20, 21]. Similar observations exist for other vector/pathogen systems, such as ticks and the causative agent of Lyme disease [22], or malaria vectors [23], in which salivary gland microbiome may also play a role [24].

3. Metabarcoding: a highly sensitive and integrative approach to disentangle vector-borne pathogen transmission cycles

NGS technologies can generate millions of sequencing reads in parallel. This massive throughput sequencing capacity can produce sequence reads from fragmented libraries of a specific genome (i.e., genome sequencing) or from a pool of PCR products. Metabarcoding approaches rely on this technology where a large number of different amplicons of taxonomic informative genes (barcodes) can be sequenced. While metagenomics refers to the identification of all genomes within a particular ecosystem or sample, metabarcoding aims to identify only a subset of them (those that are of interest for a particular question) by sequencing of millions of different amplicons of these barcodes, without a necessity for cloning (i.e., sequences are obtained directly from a mix of different amplicons of different barcodes of interest) [26].

Consequently, in the case of vector-borne pathogens, starting only from the vectors as biological samples, it is possible to target and amplify well-chosen molecular markers (barcodes) of interest with universal primer sets to identify the different actors of transmission cycles (e.g., vertebrate blood sources, midgut microbiome, pathogen diversity, and vector diversity [27]). Other ecological interactions which are not directly involved in the transmission cycles but relevant for the understanding of the vector ecology and the dynamics of the transmission cycles (e.g., plant-feeding sources, sometimes required as a source of energy for routine activities such as flight, mating, and walking or a source of protein for maturation of eggs [28]) can also be identified. A schematic representation of the metabarcoding approach for the identification of ecological interactions of disease vectors is given in **Figure 2**. After purification of the total DNA (and RNA if working with RNA pathogens) contained in each vector midgut (and salivary glands, depending on the kind of vector) **(1)**, molecular markers (barcodes) of interest are PCR amplified (after RT-PCR if working with RNA pathogens) **(2)**. Then, to identify samples, a tag/index is added to each PCR product (amplicon). The same tag is used for all the amplicons obtained from a single sample **(3)**. After high-throughput sequencing **(4)**, the millions of reads **(5)** are sorted per sample thanks to the tags added to each amplicon **(6)**.

Currently, the most common systems provide up to 384 different tags and 25 million reads per sequencing run. The depth (i.e., the number of reads or the number of sequences) obtained per molecular marker and sample depends on the number of labeled samples and the number of markers amplified per sample. For instance, if we amplify 10 molecular markers for 100 vector specimens and run at a depth of 25 million sequences, about 250,000 reads per vector specimen and 25,000 reads per marker and specimen will be theoretically obtained. This kind of multiplexing allows to considerably lower sequencing costs per sample. Downstream analyses with bioinformatics tools, such as those provided on the open access Galaxy platform [29], allows to obtain and identify the sequences corresponding to each targeted marker for each vector specimen. This approach is thus

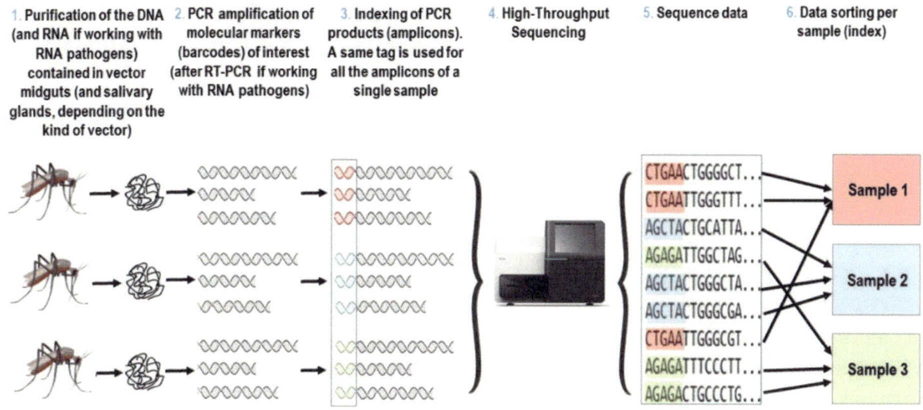

Figure 2.
Schematic representation of the metabarcoding approach for the identification of ecological interactions of disease vectors. Figure adapted from [25].

extremely powerful to further reconstitute the pathogen transmission cycles and understand its dynamics, since it can reveal, after adequate analyses, all the existing ecological interactions thanks to the simultaneous identification and for each specimen of its species or subspecies, its blood-feeding source(s), the pathogen(s) of interest, the species or lineage(s) of the pathogen(s) of interest, the composition of its midgut microbiome, of its salivary gland microbiome, its plant-feeding source(s), mutations associated with insecticide resistance, etc.

4. Unraveling *T. cruzi* transmission cycles in the Yucatan peninsula (Mexico): an example of the metabarcoding approach use

As a proof of concept, we recently performed a pilot study of the metabarcoding approach presented above using Chagas disease in the Yucatan peninsula (Mexico) [27]. In this region, *T. dimidiata* is the main vector, and different genetic subgroups of this species [30–32] live in sympatry [33]. The different molecular markers we selected for our metabarcoding approach are described below: (i) to classify *T. dimidiata* in its different genetics subgroups, we used primers targeting the Internal Transcribed Spacer ITS-2 as previously described [34]; (ii) for blood-feeding source identification, we used vertebrate universal primers targeting the 12S rRNA gene [35]; (iii) for *T. cruzi*, we used primers targeting the mini-exon gene, allowing further classification of the parasite in its different DTUs [36]; and (iv) finally, we used universal primers targeting the bacterial 16S rRNA gene to identify bacterial microbiome composition [37]. This way, we aimed to determine if there were detectable interaction patterns between the genetic subgroups of *T. dimidiata*, their blood-feeding hosts, the infection with *T. cruzi*, the parasite DTUs, and the microbiome composition, allowing elucidating at finer scales the *T. cruzi* transmission cycles in the study area.

This study, which was based on 14 *T. dimidiata* bugs collected in wild as well as in domestic ecotopes, evidences the feasibility and high sensibility of the proposed approach [27]. For example, we identified an average number of blood-feeding species per bug of 4.9 ± 0.7 and up to 7 blood-feeding species and 11 blood-feeding individuals in a single bug. Contrastingly, current techniques based on direct sequencing of PCR products can only identify the dominant sequence/host in each sample [38], while the addition of a cloning step prior to sequencing generally allows detecting up to three to five host species in some bugs [14, 39–41]. In the same way, we easily identified different DTUs infecting single bugs, while to date, most studies have relied on conventional Sanger sequencing approaches that are only capable of detecting the dominant genotype in biological samples, which almost precludes the possibility of detecting multiclonality. Based on this observation, NGS approaches capable of inventorying multiclonal infections are now being progressively adopted [42–46]. Regarding midgut microbiome, we were able to detect 23 bacterial orders and observed that its composition differed according to blood-feeding sources (**Figure 3**). Finally, all the 14 bugs belonged all to the same genetic subgroup.

To further assess potential transmission cycles of *T. cruzi* parasites by *T. dimidiata* among the identified blood source species, a feeding and parasite transmission network was constructed (**Figure 4**). Nodes of the network represent the species identified as blood meal sources, while the size of the corresponding node indicates feeding frequency on each species. Edges link species which are found together in multiple blood meals within individual bugs. Since birds cannot carry *T. cruzi* parasites, they only play a role as blood sources for triatomines, which is indicated by dotted edge connections between hosts. The solid lines between mammals indicate potential parasite transmission pathways. This network nicely highlights the mammals which would play the main role in *T. cruzi* transmission to human in the study area. Humans (*Homo sapiens* in **Figure 4**) may thus become infected by *T. cruzi* parasites originating from dogs (*Canis lupus*), cows (*Bos taurus*), and mice (*Mus musculus*), as well as from sylvatic hosts such as porcupines (*Coendou* spp.), squirrels (*Sciurus* spp.), and fruit bats (*Artibeus* spp.). Particularly, dogs appear as key actors which may favor parasite transmission to humans. This kind of networks is very informative, as it allows evidencing the animals that would play the main roles in the transmission of any pathogen to human (complementary studies focused directly on these animals may nevertheless be necessary) and that should be targeted as part of integrated control strategies aimed at disrupting parasite transmission. For example, management of the dogs and other peridomestic animals can be part of EcoHealth/One Health approaches [47]. The network presented is the result of a pilot study based on a limited sample and is only used here to illustrate the potential of the proposed metabarcoding approach. Increasing the sample size in a wide variety of ecotopes and integrating vector, microbiome and coinfection data will undoubtedly allow identifying at-risk situations and disentangling transmission cycles. It may also help to identify bacteria which are part of the normal microbiota of triatomine bugs, bacteria

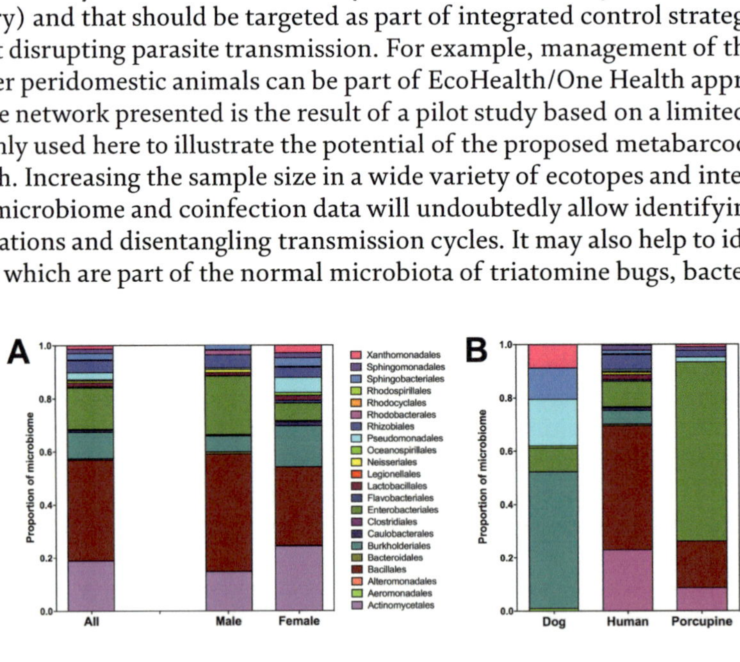

Figure 3.
Gut microbiome composition of Triatoma dimidiata. *The average composition of the microbiome from 14 individuals is shown to the level of bacterial order (A). There are significant differences between male and female microbiomes, with females presenting a greater diversity of orders. (B) Microbiome composition is also significantly different depending on the dominant blood meal present in triatomine gut, which was identified by the analysis of 12 S rRNA vertebrate sequences. Figure taken from [27].*

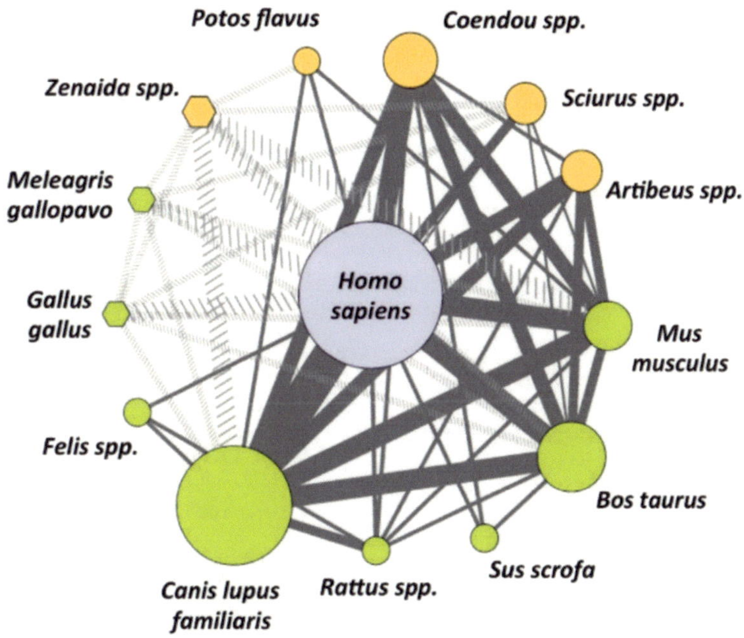

Figure 4.
Feeding and possible parasite transmission network of Triatoma dimidiata. *Blood source nodes correspond to domestic (green symbols) and sylvatic (orange symbols) host species, as well as humans (blue), with the size proportional to the feeding frequency on each host. Diamond-shaped nodes represent birds, which do not carry* Trypanosoma cruzi *parasites, and circles represent mammals, which can be infected by* T. cruzi. *Edges link species which are found together in multiple blood meals within individual bugs, and the width of the lines is proportional to the frequency of the association between species. Solid dark gray lines link mammalian species, among which* T. cruzi *may circulate, while dotted light gray lines involve bird species, which only serve as blood sources for the bugs. Humans may thus become infected by* T. cruzi *parasites originating from dogs, cows, and mice, as well as from sylvatic hosts such as porcupines, squirrels, and fruit bats. Dogs can play a key role as domestic host/reservoir favoring parasite transmission to humans. On the other hand, cats, rats, and pigs play a secondary role in parasite transmission. Figure taken from [27].*

associated with the presence/absence of infection of the bugs with *T. cruzi*, or bacteria of vital importance to the bugs. This knowledge can have important applications for the development of innovative control strategies [48–50]. The information provided by the approach can also be used to feed models including the hosts involved in the transmission to help assessing the effects of different host community managements on *T. cruzi* transmission to human and understand transmission dynamics over time [51, 52]. Transmission models are becoming increasingly important in vector-borne disease control programs. They allow evaluating different control strategies or combinations of them and assessing their cost-effectiveness and likelihood of success [53].

Consequently, the approach presented here provides very high-value information that can be used in multiple ways for further design and implementation of sustainable, effective, and locally adapted control strategies and deserves to be extended to other eco-epidemiological contexts and to any vector-borne pathogen. To date, metabarcoding approaches for the study of human vector-borne diseases using natural populations of vectors are being progressively adopted, but they are still timidly used [54, 55]. Moreover, they are still generally focused only on one of the components of transmission cycles, such as blood-feeding hosts [56–59], plant-feeding hosts [28], microbiome composition [60, 61], or vector diversity [62] (**Table 1**), thus providing limited information, while the approach can be easily more integrative, as we illustrated here, to simultaneously identify the different actors involved in transmission.

Vector	Geographic origin	Target DNA	Main findings	Reference
Mosquitoes (*Anopheles punctulatus*)	Different villages in Papua New Guinea	Mammalian blood-feeding hosts	Unbiased characterization of mammalian blood-feeding hosts, including unsuspected hosts and mixed blood meals. Human, dog, and pig were the most common host-feeding sources. The approach can also be adapted to evaluate interindividual variations among human blood meals	[56]
Mosquitoes (*Culex* and *Anopheles* spp.)	Different sites in the coast of the Caspian Sea in northern Iran	Vertebrate blood-feeding hosts	The four most common mosquito species had similar host-feeding patterns. The most commonly detected hosts in these species were humans, cattle, and ducks	[57]
Mosquitoes and sand flies	Forest sites in French Guiana	Mammal blood-feeding hosts	Accuracy of the short 12S marker proposed for the identification of Amazonian mammals. The accuracy of taxonomic assignations highly depends on the comprehensiveness of the reference library	[58]
Triatomine bugs (*Rhodnius pallescens*)	Two sampling sites in in Panama	Vertebrate blood-feeding hosts	Reliability of the metabarcoding approach proposed for the identification of vertebrate blood-feeding host	[59]
Phlebotomine sandflies (*Phlebotomus* and *Lutzomya* spp.)	Different sampling sites in Brazil, Israel, and Ethiopia	Plant-feeding hosts	Sand flies preferentially feed on *Cannabis sativa* plants. Potential utility for sand fly control	[28]
Mosquitoes (*Aedes* and *Culex* spp.)	Different habitats across central Thailand	Bacterial and eukaryotic microbiome	Patterns of microbial composition and diversity that affect pathogen prevalence appeared to differ by both vector species and habitat for a given species. Microbial composition was less diverse in urban areas	[60]
Tse-tse flies (*Glossina palpalis palpalis*)	Two trypanosomiasis foci in Cameroon	Bacterial microbiome	Endosymbiont *Wigglesworthia* was highly prominent. Potential role for *Salmonella* and *Serratia* in fly refractoriness to trypanosome infection. V4 region of the small subunit of the 16S ribosomal RNA gene was more efficient than the V3V4 region at describing the totality of the bacterial diversity	[61]
Phlebotomine sand flies (*Lutzomya* and *Brumptomyia* spp.)	Various locations in French Guiana	Sand flies	Efficiency of metabarcoding based on the mitochondrial 16S rRNA for identification of sand fly diversity in bulk samples	[62]

Vector	Geographic origin	Target DNA	Main findings	Reference
Mosquitoes and sand flies (various species)	3 sites along a gradient of anthropogenic pressure in French Guayana, area of Saint-Georges de l'Oyapock	Vectors and vertebrate blood-feeding hosts	Contrasting ecological features and feeding behavior among dipteran species, which allowed unveiling arboreal and terrestrial mammals, as well as birds, lizards, and amphibians. Lower vertebrate diversity was found in sites undergoing higher levels of human-induced perturbation	[54]
Triatomine bugs (*Triatoma dimidiata*)	Different habitats in rural Yucatan (Mexico)	Vertebrate blood-feeding hosts, *Trypanosoma cruzi* parasite, midgut bacterial microbiome, triatomine bug	Ecological associations of triatomines which shape *T. cruzi* transmission cycles. Different DTUs infecting single bugs. Identification of 14 blood-feeding species. Up to 7 blood-feeding species and 11 blood-feeding individuals identified in a single bug. Human, dog, cow, and mice were the most common host-feeding sources. Dog was highlighted as the main host involved in the pathway of *T. cruzi* transmission to human. Dynamic midgut microbiome, including 23 bacterial orders, which differed according to blood sources	[27]

Table 1.
Metabarcoding approaches for the study of human vector-borne diseases using natural populations of vectors as biological samples.

5. Conclusions

In this chapter, we presented a metabarcoding approach to study vector-borne pathogen transmission cycles and their dynamics and illustrated the feasibility and high sensitivity of the proposed approach with a recent study performed using Chagas disease in the Yucatan peninsula (Mexico), as a study model. Currently, NGS technologies are quickly becoming more affordable and cost-effective. Moreover, many bioinformatics tools have allowed to greatly simplify analyses in the last years. Consequently, this powerful approach deserves to be generalized to other eco-epidemiological contexts to unravel the transmission cycles of any vector-borne pathogen and their dynamics, which in turn will help the implementation of sustainable, effective, and locally adapted control strategies of their transmission.

Acknowledgements

This work received financial support from CONACYT (National Council of Science and Technology, Mexico) Basic Science (Project ID: CB2015-258752) and National Problems (Project ID: PN2015-893) Programs. This work was also funded by the Louisiana Board of Regents through the Board of Regents Support Fund [# LESASF (2018-2021)-RD-A-19] and grant #632083 from Tulane University School of Public Health and Tropical Medicine.

Author details

Anette Hernández-Andrade[1], Joel Moo-Millan[1], Nohemi Cigarroa-Toledo[1], Angel Ramos-Ligonio[2,3], Claudia Herrera[4], Bruno Bucheton[5], Jean-Mathieu Bart[5], Vincent Jamonneau[5], Anne-Laure Bañuls[6], Christophe Paupy[6], David Roiz[6], Denis Sereno[5], Carlos N. Ibarra-Cerdeña[7], Carlos Machaín-Williams[1], Julián García-Rejón[1], Sébastien Gourbière[8], Christian Barnabé[5], Jenny Telleria[5], Bruno Oury[5], Frédérique Brenière[5], Frédéric Simard[6], Miguel Rosado[1], Philippe Solano[5], Eric Dumonteil[4] and Etienne Waleckx[1,3,5]*

1 Centro de Investigaciones Regionales "Dr Hideyo Noguchi", Universidad Autónoma de Yucatán, Mérida, Yucatán, México

2 LADISER Inmunología y Biología Molecular, Facultad de Ciencias Químicas, Universidad Veracruzana, Orizaba, Veracruz, México

3 ACCyC, Associación Chagas con Ciencia y Conocimiento, A. C

4 Department of Tropical Medicine, School of Public Health and Tropical Medicine, Vector-Borne and Infectious Diseases Research Center, Tulane University, New Orleans, LA, USA

5 Institut de Recherche pour le Développement, UMR INTERTRYP IRD, CIRAD, Université de Montpellier, Montpellier, France

6 Institut de Recherche pour le Développement, UMR MIVEGEC, IRD, CNRS, Université de Montpellier, Montpellier, France

7 Departamento de Ecología Humana, Centro de Investigación y de Estudios Avanzados del IPN (Cinvestav), Unidad Mérida, Mérida, Yucatán, México

8 UMR5096 'Laboratoire Génome et Développement des Plantes', Université de Perpignan Via Domitia, Perpignan, France

*Address all correspondence to: etienne.waleckx@ird.fr; etienne.waleckx@correo.uady.mx

IntechOpen

© 2019 The Author(s). Licensee IntechOpen. This chapter is distributed under the terms of the Creative Commons Attribution License (http://creativecommons.org/licenses/by/3.0), which permits unrestricted use, distribution, and reproduction in any medium, provided the original work is properly cited. CC BY

References

[1] WHO. Vector-borne Diseases – Key facts [Internet]. 2017. Available from: http://www.who.int/news-room/fact-sheets/detail/vector-borne-diseases [Accessed: 06 September 2019]

[2] WHO. WHO global health days: about vector-borne diseases [Internet]. 2014. Available from: https://www.who.int/campaigns/world-health-day/2014/vector-borne-diseases/en/ [Accessed: 06 September 2019]

[3] Bhatt S, Gething PW, Brady OJ, Messina JP, Farlow AW, Moyes CL, et al. The global distribution and burden of dengue. Nature. 2013;**496**(7446):504-507. DOI: 10.1038/nature12060

[4] Pérez-Molina JA, Molina I. Chagas disease. Lancet. 2018;**391**(10115):82-94. DOI: 10.1016/s0140-6736(17)31612-4

[5] WHO. Chagas disease (American trypanosomiasis), Fact sheet N°340 [Internet]. 2019. Available from: http://www.who.int/mediacentre/factsheets/fs340/en/ [Accessed: 06 September 2019]

[6] Waleckx E, Gourbière S, Dumonteil E. Intrusive versus domiciliated triatomines and the challenge of adapting vector control practices against Chagas disease. Memórias do Instituto Oswaldo Cruz. 2015;**110**(3):324-338. DOI: 10.1590/0074-02760140409

[7] Akhoundi M, Kuhls K, Cannet A, Votýpka J, Marty P, Delaunay P, et al. A historical overview of the classification, evolution, and dispersion of *Leishmania* parasites and sandflies. PLoS Neglected Tropical Diseases. 2016;**10**(3):e0004349. DOI: 10.1371/journal.pntd.0004349

[8] Ibarra-Cerdeña CN, Valiente-Banuet L, Sanchez-Cordero V, Stephens CR, Ramsey JM. *Trypanosoma cruzi* reservoir-triatomine vector co-occurrence networks reveal meta-community effects by synanthropic mammals on geographic dispersal. PeerJ. 2017;**5**:e3152. DOI: 10.7717/peerj.3152

[9] Brenière SF, Waleckx E, Barnabé C. Over six thousand *Trypanosoma cruzi* strains classified into discrete typing units (DTUs): Attempt at an inventory. PLoS Neglected Tropical Diseases. 2016;**10**(8):e0004792. DOI: 10.1371/journal.pntd.0004792

[10] Brenière SF, Buitrago R, Waleckx E, Depickère S, Sosa V, Barnabé C, et al. Wild populations of *Triatoma infestans*: Compilation of positive sites and comparison of their ecological niche with domestic population niche. Acta Tropica. 2017;**176**:228-235. DOI: 10.1016/j.actatropica.2017.08.009

[11] Noireau F, Diosque P, Jansen AM. *Trypanosoma cruzi*: Adaptation to its vectors and its hosts. Veterinary Research. 2009;**40**(2):26. DOI: 10.1051/vetres/2009009

[12] Dorn PL, Justi S, Krafsur ES, Lanzaro GC, Cornel AJ, Lee Y, et al. Genetics of major insect vectors. In: Tibayrenc M, editor. Genetics and Evolution of Infectious Diseases. 2nd ed. London: Elsevier; 2017. pp. 341-382. DOI: 10.1016/B978-0-12-799942-5.00015-9

[13] Rabinovich JE, Kitron UD, Obed Y, Yoshioka M, Gottdenker N, Chaves LF. Ecological patterns of blood-feeding by kissing-bugs (Hemiptera: Reduviidae: Triatominae). Memórias do Instituto Oswaldo Cruz. 2011;**106**(4):479-494. DOI: 10.1590/s0074-02762011000400016

[14] Waleckx E, Suarez J, Richards B, Dorn PL. *Triatoma sanguisuga* blood meals and potential for Chagas disease, Louisiana, USA. Emerging Infectious

Diseases. 2014;**20**(12):2141-2143. DOI: 10.3201/eid2012.131576

[15] Azambuja P, Garcia ES, Ratcliffe NA. Gut microbiota and parasite transmission by insect vectors. Trends in Parasitology. 2005;**21**(12):568-572. DOI: 10.1016/j.pt.2005.09.011

[16] Weiss BL, Wang J, Maltz MA, Wu Y, Aksoy S. Trypanosome infection establishment in the tsetse fly gut is influenced by microbiome-regulated host immune barriers. PLoS Pathogens. 2013;**9**(4):e1003318. DOI: 10.1371/journal.ppat.1003318

[17] Kelly PH, Bahr SM, Serafim TD, Ajami NJ, Petrosino JF, Meneses C, et al. The gut microbiome of the vector *Lutzomyia longipalpis* is essential for survival of *Leishmania infantum*. MBio. 2017;**8**(1):e01121-16. DOI: 10.1128/mBio.01121-16

[18] Ramirez JL, Souza-Neto J, Torres Cosme R, Rovira J, Ortiz A, Pascale JM, et al. Reciprocal tripartite interactions between the *Aedes aegypti* midgut microbiota, innate immune system and dengue virus influences vector competence. PLoS Neglected Tropical Diseases. 2012;**6**(3):e1561. DOI: 10.1371/journal.pntd.0001561

[19] Hill CL, Sharma A, Shouche Y, Severson DW. Dynamics of midgut microflora and dengue virus impact on life history traits in *Aedes aegypti*. Acta Tropica. 2014;**140**:151-157. DOI: 10.1016/j.actatropica.2014.07.015

[20] Flores HA, O'Neill SL. Controlling vector-borne diseases by releasing modified mosquitoes. Nature Reviews. Microbiology. 2018;**16**(8):508-518. DOI: 10.1038/s41579-018-0025-0

[21] Jupatanakul N, Sim S, Dimopoulos G. The insect microbiome modulates vector competence for arboviruses. Viruses. 2014;**6**(11): 4294-4313. DOI: 10.3390/v6114294

[22] Narasimhan S, Rajeevan N, Liu L, Zhao YO, Heisig J, Pan J, et al. Gut microbiota of the tick vector *Ixodes scapularis* modulate colonization of the Lyme disease spirochete. Cell Host & Microbe. 2014;**15**(1):58-71. DOI: 10.1016/j.chom.2013.12.001

[23] Boissière A, Tchioffo MT, Bachar D, Abate L, Marie A, Nsango SE, et al. Midgut microbiota of the malaria mosquito vector *Anopheles gambiae* and interactions with *Plasmodium falciparum* infection. PLoS Pathogens. 2012;**8**(5):e1002742. DOI: 10.1371/journal.ppat.1002742

[24] Tchioffo MT, Boissière A, Abate L, Nsango SE, Bayibéki AN, Awono-Ambéné PH, et al. Dynamics of bacterial community composition in the malaria mosquito's epithelia. Frontiers in Microbiology. 2016;**6**:1500. DOI: 10.3389/fmicb.2015.01500

[25] Waleckx E, Arnal A, Dumonteil E. Metabarcoding, un nuevo enfoque para el estudio de los ciclos de transmisión de la enfermedad de Chagas y la ecología de sus vectores. In: Zamora-Bustillos R, Sandoval-Gío JJ, editors. Contribución de la Biotecnología al Desarrollo de la Península de Yucatán. Mérida, Yucatán, México: TECNM; 2019. pp. 1052-1065

[26] Taberlet P, Coissac E, Pompanon F, Brochmann C, Willerslev E. Towards next-generation biodiversity assessment using DNA metabarcoding. Molecular Ecology. 2012;**21**(8):2045-2050. DOI: 10.1111/j.1365-294X.2012.05470.x

[27] Dumonteil E, Ramirez-Sierra MJ, Perez-Carrillo S, Teh-Poot C, Herrera C, Gourbiere S, et al. Detailed ecological associations of triatomines revealed by metabarcoding and next-generation sequencing: Implications for triatomine behavior and *Trypanosoma cruzi* transmission cycles. Scientific Reports. 2018;**8**(1):4140. DOI: 10.1038/s41598-018-22455-x

[28] Abbasi I. Trancoso Lopo de Queiroz a, Kirstein OD, Nasereddin a, Horwitz BZ, Hailu a, et al. plant-feeding phlebotomine sand flies, vectors of leishmaniasis, prefer *Cannabis sativa*. Proceedings of the National Academy of Sciences of the United States of America. 2018;**115**(46):11790-11795. DOI: 10.1073/pnas.1810435115

[29] Kosakovsky Pond S, Wadhawan S, Chiaromonte F, Ananda G, Chung WY, Taylor J, et al. Windshield splatter analysis with the galaxy metagenomic pipeline. Genome Research. 2009;**19**(11):2144-2153. DOI: 10.1101/gr.094508.109

[30] Bargues MD, Klisiowicz DR, Gonzalez-Candelas F, Ramsey JM, Monroy C, Ponce C, et al. Phylogeography and genetic variation of *Triatoma dimidiata*, the main Chagas disease vector in Central America, and its position within the genus *Triatoma*. PLoS Neglected Tropical Diseases. 2008;**2**(5):e233. DOI: 10.1371/journal.pntd.0000233

[31] Justi SA, Cahan S, Stevens L, Monroy C, Lima-Cordon R, Dorn PL. Vectors of diversity: Genome wide diversity across the geographic range of the Chagas disease vector *Triatoma dimidiata* sensu lato (Hemiptera: Reduviidae). Molecular Phylogenetics and Evolution. 2018;**120**:144-150. DOI: 10.1016/j.ympev.2017.12.016

[32] Monteiro FA, Peretolchina T, Lazoski C, Harris K, Dotson EM, Abad-Franch F, et al. Phylogeographic pattern and extensive mitochondrial DNA divergence disclose a species complex within the Chagas disease vector *Triatoma dimidiata*. PLoS One. 2013;**8**(8):e70974. DOI: 10.1371/journal.pone.0070974

[33] Herrera-Aguilar M, Be-Barragan LA, Ramirez-Sierra MJ, Tripet F, Dorn P, Dumonteil E. Identification of a large hybrid zone between sympatric sibling species of *Triatoma dimidiata* in the Yucatan peninsula, Mexico and its epidemiological importance. Infection, Genetics and Evolution. 2009;**9**(6):1345-1351. DOI: 10.1016/j.meegid.2009.09.009

[34] Richards B, Rua NM, Monroy C, Stevens L, Dorn PL. Novel polymerase chain reaction-restriction fragment length polymorphism assay to determine internal transcribed spacer-2 group in the Chagas disease vector, *Triatoma dimidiata* (Latreille, 1811). Memórias do Instituto Oswaldo Cruz. 2013;**108**(4):395-398. DOI: 10.1590/s0074-0276108042013001

[35] Kitano T, Umetsu K, Tian W, Osawa M. Two universal primer sets for species identification among vertebrates. International Journal of Legal Medicine. 2007;**121**(5):423-427. DOI: 10.1007/s00414-006-0113-y

[36] Zingales B, Andrade SG, Briones MR, Campbell DA, Chiari E, Fernandes O, et al. A new consensus for *Trypanosoma cruzi* intraspecific nomenclature: Second revision meeting recommends TcI to TcVI. Memórias do Instituto Oswaldo Cruz. 2009;**104**:1051-1054. DOI: 10.1590/S0074-02762009000700021

[37] Baker GC, Smith JJ, Cowan DA. Review and re-analysis of domain-specific 16S primers. Journal of Microbiological Methods. 2003;**55**(3):541-555. DOI: 10.1016/j.mimet.2003.08.009

[38] Gottdenker NL, Chaves LF, Calzada JE, Saldana A, Carroll CR. Host life history strategy, species diversity, and habitat influence *Trypanosoma cruzi* vector infection in changing landscapes. PLoS Neglected Tropical Diseases. 2012;**6**(11):e1884. DOI: 10.1371/journal.pntd.0001884

[39] Gorchakov R, Trosclair LP, Wozniak EJ, Feria PT, Garcia MN, Gunter SM, et al. *Trypanosoma cruzi*

infection prevalence and bloodmeal analysis in triatomine vectors of Chagas disease from rural peridomestic locations in Texas, 2013-2014. Journal of Medical Entomology. 2016;**53**(4):911-918. DOI: 10.1093/jme/tjw040

[40] Stevens L, Monroy MC, Rodas AG, Dorn PL. Hunting, swimming, and worshiping: Human cultural practices illuminate the blood meal sources of cave dwelling Chagas vectors (*Triatoma dimidiata*) in Guatemala and Belize. PLoS Neglected Tropical Diseases. 2014;**8**(9):e3047. DOI: 10.1371/journal.pntd.0003047

[41] Moo-Millan J, Arnal A, Pérez-Carrillo S, Hernandez-Andrade A, Ramírez-Sierra MJ, Rosado-Vallado M, et al. Disentangling *Trypanosoma cruzi* transmission cycle dynamics through the identification of blood meal sources of natural populations of *Triatoma dimidiata* in Yucatán, Mexico. Parasites and Vectors. in press

[42] Pronovost H, Peterson AC, Chavez BG, Blum MJ, Dumonteil E, Herrera CP. Deep sequencing reveals multiclonality and new discrete typing units of *Trypanosoma cruzi* in rodents from the southern United States. Journal of Microbiology, Immunology, and Infection. 2018;**S1684-1182**(18):30097-30095. DOI: 10.1016/j.jmii.2018.12.004

[43] Herrera C, Majeau A, Didier P, Falkenstein KP, Dumonteil E. *Trypanosoma cruzi* diversity in naturally infected nonhuman primates in Louisiana assessed by deep sequencing of the mini-exon gene. Transactions of the Royal Society of Tropical Medicine and Hygiene. 2019;**113**(5):281-286. DOI: 10.1093/trstmh/try119

[44] Villanueva-Lizama L, Teh-Poot C, Majeau A, Herrera C, Dumonteil E. Molecular genotyping of *Trypanosoma cruzi* by next-generation sequencing of the mini-exon gene reveals infections with multiple parasite discrete typing units in chagasic patients from Yucatan, Mexico. The Journal of Infectious Diseases. 2019;**219**(12):1980-1988. DOI: 10.1093/infdis/jiz047

[45] Majeau A, Herrera C, Dumonteil E. An improved approach to *Trypanosoma cruzi* molecular genotyping by next-generation sequencing of the mini-exon gene. Methods in Molecular Biology. 1955;**2019**:47-60. DOI: 10.1007/978-1-4939-9148-8_4

[46] Llewellyn MS, Messenger LA, Luquetti AO, Garcia L, Torrico F, Tavares SB, et al. Deep sequencing of the *Trypanosoma cruzi* GP63 surface proteases reveals diversity and diversifying selection among chronic and congenital Chagas disease patients. PLoS Neglected Tropical Diseases. 2015;**9**(4):e0003458. DOI: 10.1371/journal.pntd.0003458

[47] De Urioste-Stone SM, Pennington PM, Pellecer E, Aguilar TM, Samayoa G, Perdomo HD, et al. Development of a community-based intervention for the control of Chagas disease based on peridomestic animal management: An eco-bio-social perspective. Transactions of the Royal Society of Tropical Medicine and Hygiene. 2015;**109**(2):159-167. DOI: 10.1093/trstmh/tru202

[48] Beard CB, Dotson EM, Pennington PM, Eichler S, Cordon-Rosales C, Durvasula RV. Bacterial symbiosis and paratransgenic control of vector-borne Chagas disease. International Journal for Parasitology. 2001;**31**(5-6):621-627. DOI: 10.1016/s0020-7519(01)00165-5

[49] Sassera D, Epis S, Pajoro M, Bandi C. Microbial symbiosis and the control of vector-borne pathogens in tsetse flies, human lice, and triatomine bugs. Pathogens and Global Health. 2013;**107**(6):285-292. DOI: 10.1179/2047773213y.0000000109

[50] Taracena ML, Oliveira PL, Almendares O, Umana C, Lowenberger C, Dotson EM, et al. Genetically modifying the insect gut microbiota to control Chagas disease vectors through systemic RNAi. PLoS Neglected Tropical Diseases. 2015;**9**(2):e0003358. DOI: 10.1371/journal.pntd.0003358

[51] Flores-Ferrer A, Waleckx E, Rascalou G, Dumonteil E, Gourbière S. *Trypanosoma cruzi* transmission dynamics in a synanthropic and domesticated host community. PLoS Neglected Tropical Diseases. in press

[52] Nouvellet P, Dumonteil E, Gourbiere S. The improbable transmission of *Trypanosoma cruzi* to human: The missing link in the dynamics and control of Chagas disease. PLoS Neglected Tropical Diseases. 2013;7(11):e2505. DOI: 10.1371/journal.pntd.0002505

[53] Hollingsworth TD, Adams ER, Anderson RM, Atkins K, Bartsch S, Basáñez MG, et al. Quantitative analyses and modelling to support achievement of the 2020 goals for nine neglected tropical diseases. Parasites and Vectors. 2015;**8**:630. DOI: 10.1186/s13071-015-1235-1

[54] Kocher A, de Thoisy B, Catzeflis F, Valière S, Bañuls A-L, Murienne J. iDNA screening: Disease vectors as vertebrate samplers. Molecular Ecology. 2017;**26**(22):6478-6486. DOI: 10.1111/mec.14362

[55] Titcomb GC, Jerde CL, Young HS. High-Throughput sequencing for understanding the ecology of emerging infectious diseases at the wildlife-human interface. Frontiers in Ecology and Evolution. 2019;7;126. DOI: 10.3389/fevo.2019.00126

[56] Logue K, Keven JB, Cannon MV, Reimer L, Siba P, Walker ED, et al. Unbiased characterization of *Anopheles* mosquito blood meals by targeted high-throughput sequencing. PLoS Neglected Tropical Diseases. 2016;**10**(3):e0004512. DOI: 10.1371/journal.pntd.0004512

[57] Shahhosseini N, Friedrich J, Moosa-Kazemi SH, Sedaghat MM, Kayedi MH, Tannich E, et al. Host-feeding patterns of *Culex* mosquitoes in Iran. Parasites & Vectors. 2018;**11**(1):669. DOI: 10.1186/s13071-018-3237-2

[58] Kocher A, de Thoisy B, Catzeflis F, Huguin M, Valière S, Zinger L, et al. Evaluation of short mitochondrial metabarcodes for the identification of Amazonian mammals. Methods in Ecology and Evolution. 2017;**8**(10):1276-1283. DOI: 10.1111/2041-210x.12729

[59] Kieran TJ, Gottdenker NL, Varian CP, Saldana A, Means N, Owens D, et al. Blood meal source characterization using Illumina sequencing in the chagas disease vector Rhodnius pallescens (Hemiptera: Reduviidae) in Panama. Journal of Medical Entomology. 2017;**54**(6):1786-1789. DOI: 10.1093/jme/tjx170

[60] Thongsripong P, Chandler JA, Green AB, Kittayapong P, Wilcox BA, Kapan DD, et al. Mosquito vector-associated microbiota: Metabarcoding bacteria and eukaryotic symbionts across habitat types in Thailand endemic for dengue and other arthropod-borne diseases. Ecology and Evolution. 2018;**8**(2):1352-1368. DOI: 10.1002/ece3.3676

[61] Tsagmo Ngoune JM, Reveillaud J, Sempere G, Njiokou F, Melachio TT, Abate L, et al. The composition and abundance of bacterial communities residing in the gut of *Glossina palpalis palpalis* captured in two sites of southern Cameroon. Parasites and

Vectors. 2019;**12**(1):151. DOI: 10.1186/s13071-019-3402-2

[62] Kocher A, Gantier JC, Gaborit P, Zinger L, Holota H, Valiere S, et al. Vector soup: High-throughput identification of Neotropical phlebotomine sand flies using metabarcoding. Molecular Ecology Resources. 2017;**17**(2):172-182. DOI: 10.1111/1755-0998.12556

Section 3

Pathology Diagnosis and Treatment

Chapter 8

Wnt5A Signaling Antagonizes *Leishmania donovani* Infection

Arijit Chakraborty, Shreyasi Maity and Malini Sen

Abstract

Infection by the parasite *Leishmania donovani* causes Visceral Leishmaniasis, a deadly disease if not properly treated. *L. donovani* captures the immune defense potential of host macrophages. It disrupts immune homeostasis at least partly by inducing expression of anti-inflammatory cytokines and altering the cellular cytoskeletal framework, thereby creating a niche for its own survival. While inhibition of macrophage Wnt5A signaling promotes infection by the parasite, activation of Wnt5A signaling significantly dampens infection. Experimental evidence suggests that the antagonistic effect of Wnt5A signaling on parasite infection in macrophages may be on account of its influence on the actin cytoskeleton. Importantly, the inhibitory effect of Wnt5A on infection is sustained independent of drug sensitivity and resistance. Keeping in mind that drugs used to treat Visceral Leishmaniasis are quite toxic and the parasite develops drug resistance, revamping host Wnt5A signaling may be useful for eliminating infection independent of drug sensitivity or resistance.

Keywords: Wnt5A, *L. donovani*, macrophage, cytoskeleton, Rac1, actin

1. Introduction

Leishmaniasis is a group of neglected tropical zoonotic diseases caused by intracellular obligate parasites belonging to the genus *Leishmania*. The disease is endemic to about 60 countries and may turn out life threatening if not properly treated [1]. Clinical manifestations of the disease range from cutaneous lesions (as in Cutaneous Leishmaniasis) to visceral pathologies (as in Visceral Leishmaniasis) [2]. In this review we will focus on Visceral Leishmaniasis (VL), its causative organism *Leishmania donovani* and host immune response to *L. donovani* infection.

Like many other intracellular parasites, *L. donovani* has evolved a unique ability to sustain a parasitic mode of life. In VL, *L. donovani* resides and then proliferates mainly in the spleen, the liver and the bone marrow. This can cause splenomegaly, hepatomegaly, fever and gray discoloration of the skin—hence the name "kala azar" or "black disease". Associated hematological changes include a hemoglobin level of 7–10 g/dl, leukopenia, thrombocytopenia, pancytopenia, increased ESR and erythroid hyperplasia in the bone marrow and varying degrees of erythrophagocytosis, leukophagocytosis and granulomatous reactions. In some cases clotting abnormalities and deranged platelet function are also observed [3]. Few studies, furthermore, indicate inflammatory changes in the liver, which include hypertrophy and hyperplasia of Kupffer cells, diffused intralobular fibrosis, fibrosis of the wall of the central vein, and pericellular fibrosis [4].

IntechOpen

L. donovani invades mostly host macrophages and also dendritic cells and neutrophils, hijacking the cellular machinery for its survival through adopting various strategies to evade immune response [5]. One of the strategies employed by the parasite is to inhibit phagolysosomal maturation by altering host cell lysosomal integrity [6]. *L. donovani* also alters host cytoskeletal dynamics [7, 8]. While various host and parasite specific factors are indicated in the regulation of host phagolysosomal maturation and cytoskeletal dynamics, how such regulation occurs during host pathogen interactions is unclear and a subject of intense study.

The first and the only line of treatment for Visceral Leishmaniasis is the use of chemotherapeutic agents like the pentavalent antimonials, amphotericin and paromomycin [9–11]. Although these drugs are effective for the treatment of *L. donovani* infection, drug administration is associated with serious side effects and other issues. First, these drugs are cytotoxic in nature thus causing serious damage to the hepatic health of the ailing patients [11–13]. Secondly the *L. donovani* parasites evolve drug resistant phenotypes decreasing drug efficacy. Lastly the drugs come at a high cost thereby increasing the economic burden of the already economically challenged individuals. To counteract these challenges researchers all over the world have tried to bring out effective vaccine candidates to counteract the onset of infection and progression of the disease. Both live attenuated parasites and recombinant antigens have been used as target candidates for vaccination. Some of the vaccine candidates are still in clinical trials and their efficacy waits to be tested [14].

In spite of the use of different treatment regimens for attenuating *L. donovani* infection, understanding of one's natural host immune response to *L. donovani* infection is important. Cells of the innate arm of the immune system, for example macrophages, neutrophils and dendritic cells play an integral part in clearance of the parasites. Neutrophils have been suggested to be recruited early during the course of infection [15]. The ability of the neutrophils to produce NETs (Nuclear Extracellular Traps) and oxidative burst often determines the progression of the disease. It has been reported that neutrophils from VL patients show dysregulated NET and ROS (Reactive Oxygen Species) production. Since neutrophils are short lived the next line of defense is taken up by the macrophages. Within the macrophages the parasite tries to subvert immune defense to create a favorable niche for itself. Dendritic cells have a unique role in sustaining immune response against *L. donovani* infection. During infection with the parasite dendritic cells produce IL-12p70, a key cytokine involved in priming and maintenance of microbicidal Th1 responses [16–19]. Several reports also indicate the importance of the complement system in immune defense. The complement system involves a large number of distinct thermolabile proteins, which react with one another to opsonize pathogens and trigger a series of inflammatory responses to combat infection [20]. It has been reported that host cell complement receptors like CR3, CR1, mannose receptors (MR) are responsible for internalization of *Leishmania* [21]. In 1912, W.S. Patton first observed fresh serum mediated lysis of *L. donovani*, suggesting the role of complement system in immune defense against *L. donovani*. Later studies suggest that *L. donovani* can activate complement via the classical and alternative pathways [22]. Complement receptors have also been shown to be involved during maturation of *L. donovani* containing phagosomes [23]. In light of the genetic studies carried out on mice, gene loci *Lsh* and *H2* may be linked to resistance toward *L. donovani* infection. Gene products like Nramp1 from the *Lsh* locus of chromosome 1 in mice influence natural resistance to *L. donovani*. The H2 locus which codes for the MHC (Major Histocompatibility Complex) molecules in mice also determines the fate of the disease in mice [24]. Other gene products involved in macrophage mediated immune defense, for example *Ifn*, *Tnf* and *Nos* play a very important role in clearance of the parasite. Upregulation of these gene products however, does not limit the disease, suggesting that many other factors may be involved in the interplay of events that bolster immunity to infection [25–27]. Wnt signaling, which is known to regulate different aspects of cell polarity and coordination in tissue patterning during development is an important theme to study in this context [28–30], especially in view of the fact that different levels of cell polarity and coordination are intrinsic to the progression of innate immune responses [31].

Wnts comprise a family of about nineteen glycoprotein ligands that signal upon binding to the Frizzed (about twelve in number) and ROR1/ROR2 transmembrane cell surface receptors (**Figure 1**). While the Frizzled receptors bear resemblance to heterotrimeric G protein coupled receptors, ROR1/ROR2 bear tyrosine kinase like motifs [32]. Wnt signaling in general is divided into two broad categories, -canonical or β-catenin dependent and non-canonical or β-catenin independent. The transcriptional co-activator β-catenin promotes gene expression by LEF/TCF family transcription factors in response to canonical Wnt signaling and transcriptional activators such as NFκB, NFAT and AP1 are associated with non-canonical Wnt signaling. Although the ligands Wnt3A and Wnt5A are mostly considered as representatives of the canonical and noncanonical modes of Wnt signaling respectively, the mode of signaling is in reality governed by the receptor(s) receiving the Wnt signal and the associated adaptor molecule(s) transmitting it [33, 34]. Thus some

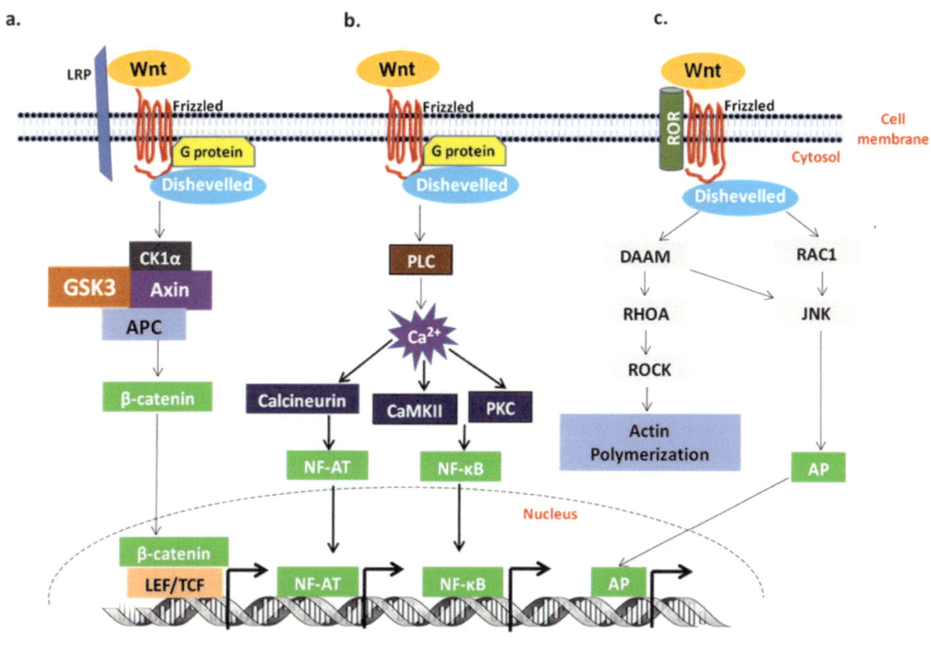

Figure 1.
Overview of Wnt signaling pathways. (a) β-Catenin dependent (the canonical pathway), (b) Wnt-Ca^{2+}-dependent pathway and (c) planar cell polarity pathway (PCP) (noncanonical pathways). (a) Wnt ligand binds to the Frizzled (Fz) family transmembrane receptors and coreceptor LRP leading to the activation of the signaling pathway through Disheveled (Dvl), Casein Kinase 1α (CK1α) and/or G proteins. This leads to inhibition of the destructive complex formation by GSK3, Axin and APC causing the stabilization, accumulation and subsequent nuclear translocation of β-catenin. β-Catenin forms an active complex with nuclear LEF and TCF promoting expression of LEF/TCF responsive genes. In the absence of Wnt ligands, APC complex can degrade β-catenin. (b) In noncanonical Wnt-Ca^{2+}-dependent pathway, Wnt-Fz interaction activates PLC via Disheveled and/or G-protein, which triggers the release of calcium ion. Increased level of calcium ion in turn activates calcineurin, CaMKII and PKC that help in nuclear translocation of NF-AT and NF-κB. (c) In noncanonical PCP pathway, Wnt may bind with ROR and Fz, which then activates Dvl. Activated Dvl can modulate actin cytoskeleton through DAAM, RHOA and ROCK signaling. Dvl also activates Rac1, which triggers JNK activity leading to nuclear translocation of AP and gene expression. LRP, low density lipoprotein receptor-related protein; GSK3 , glycogen synthase kinase 3; APC, adenomatous polyposis coli; LEF, lymphoid enhancer binding factor; TCF, T cell factor.

level of overlap between the two modes of signaling is quite frequent. Interestingly, the intracellular adaptor molecule Disheveled acts as a mediator of both β-catenin-dependent and β-catenin independent Wnt signaling. Heterotrimeric G proteins, which have been reported to couple with Frizzled receptors, add to the complexity of Wnt signaling [35, 36]. Whether heterotrimeric G proteins cooperate with Disheveled during canonical and non-canonical Wnt signaling is unclear. Despite some evidence of the involvement of lipid molecules such as cholesterol in switching Disheveled between the canonical and noncanonical modes of Wnt signaling [36], the molecular details of such presumed conformational switches remain undocumented.

In this chapter we will focus on the role of Wnt5A signaling in host immune response and its influence on *L. donovani* infection. Our present knowledge in the field of host parasite interaction limits us to the use of chemotherapeutic intervention to limit *L. donovani* infection. Therefore it is necessary to delve deep into understanding the cell biology of infection in the context of immune modulators like Wnt5A. Quite justifiably this kind of study will not only help us to understand host pathogen interaction in a better way but also aid in the formulation of novel therapeutic strategies through regulation of immune response.

2. Role of Wnt5A signaling in immune response

Primary studies on the association of Wnt5A signaling, a prototype for the non-canonical mode of Wnt signaling with immune response were focused on synovial fibroblasts from RA patients, where a correlation between Wnt5A signaling and proinflammatory cytokine expression was established [37]. Subsequent experimental evidence suggested that Wnt5A signaling may regulate IL-12 and IFN-γ expression by macrophages in the context of mycobacterial infection [38]. Similar studies carried out in bone marrow stromal cells and synovial fibroblasts also suggested that Wnt5A activates secretion of cytokines (IL-6, IL-1β) and chemokines (CCL2, CCL5, CXCL1 and CXCL5) upon interaction with Frizzled-5 and ROR, putative cell surface receptors to Wnt5A through NFκB activation [39].

It is now known that Wnt5A signaling plays an important role in stimulating microbial phagocytosis and sustenance of immune homeostasis through alteration in actin assembly and maintenance of an appropriate cytokine milieu [40–43]. The role of WNT5A signaling in stimulating microbial phagocytosis has been shown to be dependent on Rac GTPase and Rho GTPase activity in macrophages [41]. These GTPases are known to regulate cytoskeletal changes. Immune cells like macrophages, dendritic cells and other antigen presenting cells depend heavily on the cytoskeletal modulators to phagocytose and process various antigens so that they can be effectively presented to the T cells [44]. Alteration of cytoskeletal dynamics by Wnt5A signaling therein may be correlated with better assembly of NADPH oxidase subunits on phagosomal membranes and efficient production of microbicidal ROS [45, 46]. The increase in phagocytic activity and microbial killing by Wnt5A signaling may also be dependent upon the change in lipid raft organization of the macrophages in association with actin assembly [41–43]. Suppression of Wnt5A production by IWP2 (Inhibitor of Wnt production-2) accordingly abrogates both phagocytic activity and microbial killing in macrophages [40, 41, 47].

Several lines of evidence indicate that Wnt5A signaling is important for macrophage differentiation and survival. When stimulated with granulocyte monocyte-colony stimulator factor (GM-CSF), bone marrow from Wnt5A conditional knock-out mice show less potency to differentiate into $F4/80^{+}$ and $CD11b^{+}$ macrophages compared to bone marrow from control mice [48]. Furthermore,

Wnt5A-depleted BMDMs (Bone Marrow Derived Macrophages) show reduced expression of the anti-apoptotic molecules Bcl2 and Bcl-xl, and increased expression of the pro-apoptotic molecule Bax, leading to decreased survival of macrophages [40].

Depletion of steady state Wnt5A signaling reduces IFN-β and IFN-γ production by macrophages through inhibition of IκB kinase β (IKK2) activity. The reduction in IKK2 activity, which causes reduced IκB degradation and p65 (NFκB) nuclear translocation relates to decreased expression of immune regulators such as CD14 that are key components of immune responses during microbial infection [40]. Other reports suggest that depletion of Wnt5A signaling correlates with increase in microbial infections in mice [40, 43]. Thus Wnt5A signaling may be a crucial player in the sustenance of immune functions.

3. Intracellular life of *L. donovani*: the role of Wnt5A signaling therein

Intracellular parasitism is a strategy by which parasites build a niche to sustain within the host. Parasites such as *L. donovani* have developed sophisticated strategies to counteract host defense machinery. One such strategy to adapt to a parasitic mode of life is the dimorphic life cycle in *L. donovani*. *L. donovani* resides in the gut of its vector (Phlebotomine sand flies) in a flagellated infective form (promastigote). During a blood meal of the sand fly on its mammalian host these promastigotes are transferred to the blood, where they are phagocytosed by neutrophils and macrophages. While residing in macrophages, the parasites lose their flagella and transform to amastigotes. Amastigotes divide and thrive within the host, causing disease. The parasite life cycle is repeated with blood meal of sandflies from parasite-infected patients. The mechanism of entry of *L. donovani* into macrophages has been debated for long. It has been shown that host cell receptors (for example Complement receptors and Fcγ) influence *L. donovani* internalization and this interaction is partially dependent on the presence of promastigote flagella [49]. It is also documented that host cell membrane microdomains influence internalization of the parasite [5, 7, 50] . In order to hijack the cellular defense machinery *L. donovani* interacts with components of endoplasmic reticulum and the trans-Golgi network (TGN) [7, 51]. *L. donovani* containing vacuoles take up necessary nutrients like glucose, amino acid and essential ions like Fe^{2+} from the trans-Golgi network (TGN). These parasite harboring vacuoles/parasitophorous vacuoles (PV) while acquiring nutrients also disrupt the transport of different proteins to their designated vacuoles (endosomes/lysosomes) from the TGN and endoplasmic reticulum (ER), thus compromising their function [51]. The internalized parasite also delays the fusion of PV with the lysosomes through the action of lipophosphoglycan (LPG), a parasite derived molecule. Parasitophorous vacuoles accordingly become encapsulated with host F-actin, myosin and F-actin nucleating factors, thus producing a halo of F-actin surrounding the vacuole and inhibiting its lysosomal fusion [6]. The parasitophorous vacuole also expresses the early endosomal marker EEA1, and the small GTPases Rab5 and Rab 7 [52] preventing lysosomal degradation. The altered acidification of parasitophorous vacuoles is instrumental in promastigote to amastigote transformation and sustenance of infection [53]. Such remodeling of PV may lead to alteration in host lipid microdomains and alter assembly of the NADPH oxidase complex, which holds a key to microbial elimination through generation of microbiocidal Reactive Oxygen Species (ROS). The influence of Wnt5A signaling on actin cytoskeletal dynamics, organization of lipid raft microdomains and organelle polarity and assembly [30, 40, 41] suggests that host macrophages can potentially counteract the establishment and progression of *L. donovani* infection through Wnt5A signaling.

L. donovani infection is accompanied by increase in anti-inflammatory cytokine expression, which may help the intracellular amastigotes to build a safe niche within the macrophage. Increase in anti-inflammatory cytokines is often associated with decrease in production of ROS or Nitric oxide, which is unfavorable for amastigote growth [54–56]. Host macrophage Wnt5A signaling may be instrumental in attenuating the effect of anti-inflammatory cytokines by maintaining a proinflammatory cytokine signature [37, 40, 41].

4. Host defense against *Leishmania donovani* infection in the context of Wnt5A signaling

Macrophages, the primary sentinels of host immune response carry the potential to confront the challenge imposed by *L. donovani* infection. One important strategy adopted by macrophages to limit the pathogenicity from infection is production of ROS and nitric oxide, which are detrimental to the pathogen [57]. Interestingly the production of ROS and nitric oxide is often triggered by cytokine or chemokine signaling in the macrophages. IFN-γ is considered to be one of the major chemokines which bring about production of nitric oxide and ROS in macrophages [58]. IFN-γ null and IFN-γ receptor null mice carry enhanced infection load [50] substantiating the importance of IFN-γ in restraining infection. Overall, cytoskeletal actin modulations in association with organization of protein-lipid microdomains and transcriptional control of immune regulatory genes act in concert to antagonize the attempts of the parasite to settle into a favored niche within macrophages. These aspects of immune response to parasite infection are akin to the already known attributes of Wnt5A signaling as described previously in this chapter. Thus keeping in mind that *L. donovani* tries to subvert immune response by modulating lipid dynamics as well as cytoskeletal dynamics it is important to study the role of Wnt5A signaling in *L. donovani* infection.

Experimental evidence indicates that Wnt5A signaling and *L. donovani* infection are in mutual opposition. *In-vitro studies* have shown that the protein level of Wnt5A in *L. donovani* infected macrophages is significantly lower than that in the uninfected controls, with no significant change in Wnt5A mRNA level. The observed decrease in Wnt5A protein upon *L. donovani* infection is indicative of infection-induced suppression of Wnt5A signaling in the host macrophage. Protozoan parasites like *L. donovani,* are known to harbor a plethora of proteases (including metalloproteases) to counteract host immune response through cleavage of host proteins. GP63, one such well-studied metalloprotease, is expressed in significant amounts in both the promastigote and amastigote forms of the *L. donovani* parasite. Cleavage and destruction of host proteins such as AP1 by GP63 has already been reported [54]. Since, the reduction in Wnt5A protein upon infection of macrophages with *L. donovani* was inhibited by O-phenanthroline, a small molecule inhibitor of metalloproteases, it is possible that infection induced reduction in Wnt5A protein is brought about by the action of parasite specific metalloproteases like GP63. In view of the fact that Wnt5A signaling is known to boost immune homeostasis [40, 41], reduction in Wnt5A protein level may help the parasite to evade immune response. In contrary to our observation, the mRNA levels of Wnt5A are found to increase during mycobacterial and ehrlichial infection in macrophages [38, 59]. Changes in Wnt5A mRNA and protein levels may depend on the type and load of infection. Thus, further validation of proteomic data from various samples such as sera and spleen aspirates from *L. donovani* infected individuals will be needed to further analyze and understand the experimental findings.

On the basis of the understanding that the steady state level of Wnt5A signaling is significantly reduced during *L. donovani* infection, we hypothesized that revamping Wnt5A signaling in macrophages might have a debilitating effect on parasite load. Indeed we found that upon treating macrophages with recombinant Wnt5A prior to infection there is significant reduction in parasite load. The decrease in the parasite load was seen to be dose and time dependent. Interestingly, a decrease in parasite load was also seen when Wnt5A was exogenously added to infected macrophages, suggesting that there may be a therapeutic role of Wnt5A signaling during *L. donovani* infection [42]. Depletion of Wnt5A signaling through application of IWP-2 or transfection with Frizzled5 (putative receptor to Wnt5A) siRNA resulted in enhancement of infection by *L. donovani*, corroborating the importance of Wnt5A signaling in limiting *L. donovani* infection.

There is evidence that Wnt5A signaling mediated killing of *L. donovani* within the macrophages is brought about by change in Wnt5A induced cytoskeletal and membrane dynamics. Revamping host Wnt5A signaling by exogenous Wnt5A leads to reduction in parasite survival probably because the lay out for a self-sustaining parasite niche in the form of parasitophorous vacuole (PV) is not compatible with the cytoskeletal alterations and associated endosomal/lysosomal vesicle movements induced by Wnt5A. Enhanced endolysosomal fusion in infected macrophages occurred through Wnt5A signaling in infected cells as judged by live cell microscopy

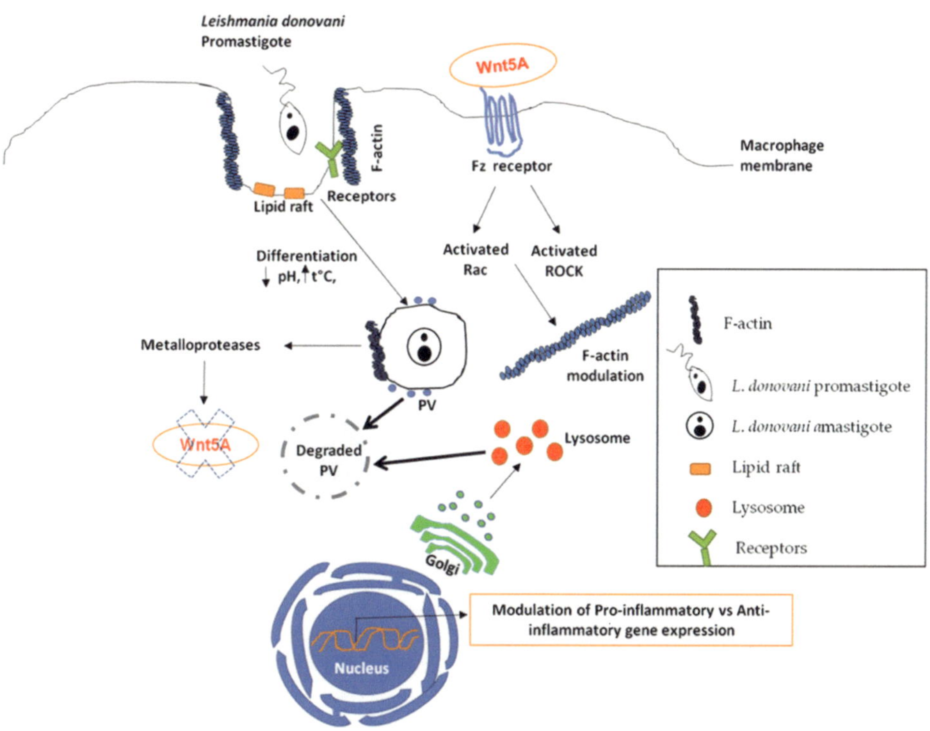

Figure 2.
Schematic representation of Wnt5A signaling-mediated inhibition of Leishmania donovani *infection.* Leishmania *promastigote interacts with the macrophage through different macrophage receptors. This interaction leads to the formation of parasite containing early endosome where several factors like low pH and increased temperature help in differentiation of the parasite into amastigote form. It secretes metalloproteases, which can degrade Wnt5A. Activated Wnt5A signaling modulates actin cytoskeleton and promotes phagolysosomal fusion leading to degradation of the parasite. Wnt5A signaling may also alter the expression of pro-inflammatory and anti-inflammatory cytokine genes thus promoting parasite clearance. PV, parasitophorous vacuole; ROCK, rho-associated coiled coil kinase; Rac, Ras-related C3 botulinum toxin substrate.*

using live cell dyes [42]. Electron microscopic (EM) images showed Wnt5A induced increased membranous vesicle fusions with PV in the infected cells. The EM images also revealed a low abundance of the PV upon Wnt5A treatment. The apparent membranous wrappings in degraded PV, as suggested by EM may be due to the formation of autophagosomes through fusion of PV with membranous aggregates during cytoskeletal movements [42]. One of the strategies that a host may adopt through Wnt5A signaling is laminate the parasitophorous vacuole so that the solutes cannot easily reach the parasite thereby slowly starving the parasite to death. While laminating the parasitophorous vacuole it may also ensure that the NADPH oxidase is well assembled so as to generate adequate amounts of ROS, which could lead to killing of parasites. The membrane lamination on the parasitophorous vacuoles through enhanced cytoskeletal dynamics could also lead to increased PV-lysosomal fusion thereby promoting rapid degradation of the parasite. Our study demonstrates that Wnt5A signaling mediated killing of *L. donovani* in macrophages is abrogated when inhibitors of cytoskeletal proteins like Rac1 GTPase and Rho kinase are used, thus implying that the effects of Wnt5A signaling on infection are at least partly mediated through the small molecular weight actin associated GTPases. The possible mechanism of Wnt5A signaling mediated parasite clearance is depicted in **Figure 2**.

It will be important to validate the effect of Wnt5A signaling on *L. donovani* infection *in vivo* and also check the load of *L. donovani* infection in Wnt5A heterozygous mice (Wnt5A null are lethal). Analysis of the cytokine milieu *in vivo* upon activation of Wnt5A signaling at the onset of infection will also provide useful information about the mechanism of Wnt5A induced containment of infection.

5. Conclusion

Wnt5A signaling maintains immune homeostasis [40]. If Wnt5A signaling is not sufficient a disturbed immune homeostasis could lead to adverse effects during *L. donovani* infection. Recently, it has been suggested that *L. donovani* infection associated with a skewed hematopoiesis program promotes the visceral disease [60]. Since Wnt5A signaling is involved in hematopoiesis [61], it is important to have a clear understanding of the role of Wnt5A directed hematopoiesis during *L. donovani* infection.

L. donovani through its evolution has undergone various changes to accommodate itself efficiently in its ever-changing environment. Often drugs have been rendered useless by the emergence of drug resistant strains. Therefore, it would be an efficient strategy to identify host cell factors, which act against these infections and revamp them. Our results indicate that host Wnt5A signaling restricts infection by both antimony drug sensitive and resistant *L. donovani* strains at least partly by prohibiting parasite niche formation within host macrophages. Interestingly, in a follow up study we found a similar kind of result with *Pseudomonas aeruginosa* or *Streptococcus pneumoniae*. These pathogenic bacteria degrade Wnt5A from the system and when Wnt5A is added exogenously the macrophages efficiently lower the bacterial load. The clearance of bacteria was found to happen through cytoskeletal reorganization and efficient formation of LC3B containing phagosome [43]. Recently high serum levels of the Wnt antagonist Dkk1 has been correlated with a predominantly Th2 phenotype during the onset of experimental cutaneous leishmaniasis [62]. Since Wnt5A supports a pro-inflammatory cytokine signature its potential for protection against parasite infection may also involve prevention of the predominantly Th2 signature that sustains infection. Thus Wnt5A signaling plays an important role in maintaining an innate immune readiness within macrophages for pathogenic onslaught.

Acknowledgements

This work was supported by ICMR, Govt. of India (2016-0222/CMB/ADHOC-BMS dated 04/12/2018), Institutional funding (BSC0114, BSC0116). SM was supported by Research Scholar Fellowship from UGC, Govt. of India.

Conflict of interest

The authors declare that there is no conflict of interest.

Author details

Arijit Chakraborty, Shreyasi Maity and Malini Sen*
Division of Cancer Biology and Inflammatory Disorder, CSIR-Indian Institute of Chemical Biology, Kolkata, India

*Address all correspondence to: msen648@gmail.com

IntechOpen

© 2019 The Author(s). Licensee IntechOpen. This chapter is distributed under the terms of the Creative Commons Attribution License (http://creativecommons.org/licenses/by/3.0), which permits unrestricted use, distribution, and reproduction in any medium, provided the original work is properly cited.

References

[1] Pavli A, Maltezou HC. Leishmaniasis, an emerging infection in travelers. International Journal of Infectious Diseases: IJID: Official Publication of the International Society for Infectious Diseases. 2010;**14**(12):e1032-e1039

[2] Burza S, Croft SL, Boelaert M. Leishmaniasis. Lancet (London, England). 2018;**392**(10151):951-970

[3] Varma N, Naseem S. Hematologic changes in visceral leishmaniasis/kala azar. Indian Journal of Hematology & Blood Transfusion is the Official Publication of The Indian Society of Hematology & Blood Transfusion. 2010;**26**(3):78-82

[4] el Hag IA, Hashim FA, el Toum IA, Homeida M, el Kalifa M, el Hassan AM. Liver morphology and function in visceral leishmaniasis (Kala-azar). Journal of Clinical Pathology. 1994;**47**(6):547-551

[5] Liévin-Le Moal V, Loiseau PM. Leishmania hijacking of the macrophage intracellular compartments. The FEBS Journal. 2016;**283**(4):598-607

[6] Lodge R, Descoteaux A. Leishmania invasion and phagosome biogenesis. Sub-Cellular Biochemistry. 2008;**47**:174-181

[7] Moradin N, Descoteaux A. Leishmania promastigotes: Building a safe niche within macrophages. Frontiers in Cellular and Infection Microbiology. 2012;**2**:121

[8] null HA, Tejle K, Magnusson KE, Descoteaux A, Rasmusson B. Leishmania donovani lipophosphoglycan causes periphagosomal actin accumulation: Correlation with impaired translocation of PKCalpha and defective phagosome maturation. Cellular Microbiology. 2001;**3**(7):439-447

[9] Brahmachari UN. Chemotherapy of antimonial compounds in kala-azar infection. Part IV. Further observations on the therapeutic values of urea stibamine. By U.N. Brahmachari, 1922. The Indian Journal of Medical Research. 1989;**89**:393-404

[10] Croft SL, Olliaro P. Leishmaniasis chemotherapy—Challenges and opportunities. Clinical Microbiology and Infection: The Official Publication of the European Society of Clinical Microbiology and Infectious Diseases. 2011;**17**(10):1478-1483

[11] Sundar S, Rai M. Advances in the treatment of leishmaniasis. Current Opinion in Infectious Diseases. 2002;**15**(6):593-598

[12] Kato KC, Morais-Teixeira E, Reis PG, Silva-Barcellos NM, Salaün P, Campos PP, et al. Hepatotoxicity of pentavalent antimonial drug: Possible role of residual Sb(III) and protective effect of ascorbic acid. Antimicrobial Agents and Chemotherapy. 2014;**58**(1):481-488

[13] Sangshetti JN, Khan FAK, Kulkarni AA, Arote R, Patil RH. Antileishmanial drug discovery: Comprehensive review of the last 10 years. RSC Advances. 2015;**5**(41):32376-32415

[14] Didwania N, Shadab M, Sabur A, Ali N. Alternative to chemotherapy—The unmet demand against leishmaniasis. Frontiers in Immunology. 2017;**8**:1779

[15] Martínez-López M, Soto M, Iborra S, Sancho D. Leishmania hijacks myeloid cells for immune escape. Frontiers in Microbiology. 2018;**9**:883

[16] Gorak PM, Engwerda CR, Kaye PM. Dendritic cells, but not macrophages, produce IL-12 immediately following Leishmania donovani infection. European Journal of Immunology. 1998;**28**(2):687-695

[17] von Stebut E, Tenzer S. Cutaneous leishmaniasis: Distinct functions of dendritic cells and macrophages in the interaction of the host immune system with Leishmania major. International Journal of Medical Microbiology: IJMM. 2018;**308**(1):206-214

[18] Marovich MA, McDowell MA, Thomas EK, Nutman TB. IL-12p70 production by Leishmania major-harboring human dendritic cells is a CD40/CD40 ligand-dependent process. Journal of Immunology (Baltimore, MD 1950). 2000;**164**(11):5858-5865

[19] León B, López-Bravo M, Ardavín C. Monocyte-derived dendritic cells formed at the infection site control the induction of protective T helper 1 responses against Leishmania. Immunity. 2007;**26**(4):519-531

[20] Janeway CA Jr, Travers P, Walport M, Shlomchik MJ. Immunobiology. 5th ed. US, New York, NY: Garland Science; 2001

[21] Blackwell JM, Ezekowitz RA, Roberts MB, Channon JY, Sim RB, Gordon S. Macrophage complement and lectin-like receptors bind Leishmania in the absence of serum. The Journal of Experimental Medicine. 1985;**162**(1):324

[22] Brittingham A, Mosser DM. Exploitation of the complement system by Leishmania promastigotes. Parasitology Today. 1996;**12**(11):444-447

[23] Polando R, Dixit UG, Carter CR, Jones B, Whitcomb JP, Ballhorn W, et al. The roles of complement receptor 3 and Fcγ receptors during Leishmania phagosome maturation. Journal of Leukocyte Biology. 2013;**93**(6):921-932

[24] Leclercq V, Lebastard M, Belkaid Y, Louis J, Milon G. The outcome of the parasitic process initiated by Leishmania infantum in laboratory mice: A tissue-dependent pattern controlled by the Lsh and MHC loci. Journal of Immunology (Baltimore, MD 1950). 1996;**157**(10):4537-4545

[25] Gradoni L, Ascenzi P. Nitric oxide and anti-protozoan chemotherapy. Parassitologia. 2004;**46**(1-2):101-103

[26] Liew FY, Li Y, Millott S. Tumour necrosis factor (TNF-alpha) in leishmaniasis. II. TNF-alpha-induced macrophage leishmanicidal activity is mediated by nitric oxide from L-arginine. Immunology. 1990;**71**(4):556-559

[27] Mohamed HS, Ibrahim ME, Miller EN, White JK, Cordell HJ, Howson JMM, et al. SLC11A1 (formerly NRAMP1) and susceptibility to visceral leishmaniasis in the Sudan. European Journal of Human Genetics. 2004;**12**(1):66-74

[28] Qian D, Jones C, Rzadzinska A, Mark S, Zhang X, Steel KP, et al. Wnt5a functions in planar cell polarity regulation in mice. Developmental Biology. 2007;**306**(1):121-133

[29] Nusse R. Wnt Signaling. Cold Spring Harbor Perspectives in Biology. 2012;**4**(5):a011163. Available from: https://www.ncbi.nlm.nih.gov/pmc/articles/PMC3331700/

[30] Witze ES, Litman ES, Argast GM, Moon RT, Ahn NG. Wnt5a control of cell polarity and directional movement by polarized redistribution of adhesion receptors. Science. 2008;**320**(5874):365-369

[31] Tran CS, Eran Y, Ruch TR, Bryant DM, Datta A, Brakeman P, et al. Host cell polarity proteins participate in innate immunity to *Pseudomonas*

aeruginosa infection. Cell Host & Microbe. 2014;**15**(5):636-643

[32] Malbon CC. Frizzleds: New members of the superfamily of G-protein-coupled receptors. Frontiers in Bioscience: A Virtual Library of Medicine. 2004;**9**:1048-1058

[33] Grumolato L, Liu G, Mong P, Mudbhary R, Biswas R, Arroyave R, et al. Canonical and noncanonical Wnts use a common mechanism to activate completely unrelated coreceptors. Genes & Development. 2010;**24**(22):2517-2530

[34] The Wnt Homepage [Internet]. 2019. Available from: http://web.stanford.edu/group/nusselab/cgi-bin/wnt/

[35] Schulte G, Bryja V. The frizzled family of unconventional G-protein-coupled receptors. Trends in Pharmacological Sciences. 2007;**28**(10):518-525

[36] Aznar N, Ear J, Dunkel Y, Sun N, Satterfield K, He F, et al. Convergence of Wnt, growth factor, and heterotrimeric G protein signals on the guanine nucleotide exchange factor Daple. Science Signaling. 2018;**11**(519):eaao4220

[37] Sen M, Lauterbach K, El-Gabalawy H, Firestein GS, Corr M, Carson DA. Expression and function of wingless and frizzled homologs in rheumatoid arthritis. Proceedings of the National Academy of Sciences. 2000;**97**(6):2791-2796

[38] Blumenthal A, Ehlers S, Lauber J, Buer J, Lange C, Goldmann T, et al. The Wingless homolog WNT5A and its receptor Frizzled-5 regulate inflammatory responses of human mononuclear cells induced by microbial stimulation. Blood. 2006;**108**(3):965-973

[39] Rauner M, Stein N, Winzer M, Goettsch C, Zwerina J, Schett G, et al. WNT5A is induced by inflammatory mediators in bone marrow stromal cells and regulates cytokine and chemokine production. Journal of Bone and Mineral Research: The Official Journal of the American Society for Bone and Mineral Research. 2012;**27**(3):575-585

[40] Naskar D, Maiti G, Chakraborty A, Roy A, Chattopadhyay D, Sen M. Wnt5a-Rac1-NF-κB homeostatic circuitry sustains innate immune functions in macrophages. Journal of Immunology (Baltimore, MD 1950). 2014;**192**(9):4386-4397

[41] Maiti G, Naskar D, Sen M. The Wingless homolog Wnt5a stimulates phagocytosis but not bacterial killing. Proceedings of the National Academy of Sciences of the United States of America. 2012;**109**(41):16600-16605

[42] Chakraborty A, Kurati SP, Mahata SK, Sundar S, Roy S, Sen M. Wnt5a signaling promotes host defense against *Leishmania donovani* infection. Journal of Immunology (Baltimore, MD 1950). 2017;**199**(3):992-1002

[43] Jati S, Kundu S, Chakraborty A, Mahata SK, Nizet V, Sen M. Wnt5A signaling promotes defense against bacterial pathogens by activating a host autophagy circuit. Frontiers in Immunology. 2018;**9**:679

[44] Wickramarachchi DC, Theofilopoulos AN, Kono DH. Immune pathology associated with altered actin cytoskeleton regulation. Autoimmunity. 2010;**43**(1):64-75

[45] Bokoch GM, Zhao T. Regulation of the phagocyte NADPH oxidase by Rac GTPase. Antioxidants & Redox Signaling. 2006;**8**(9-10):1533-1548

[46] Pendyala S, Usatyuk PV, Gorshkova IA, Garcia JG, Natarajan V. Regulation of NADPH oxidase in vascular endothelium: The role of phospholipases, protein kinases, and

cytoskeletal proteins. Antioxidants & Redox Signaling. 2019;**11**(4):841-860. Available from: https://www.ncbi.nlm.nih.gov/pubmed/18828698

[47] Chen B, Dodge ME, Tang W, Lu J, Ma Z, Fan CW, et al. Small molecule-mediated disruption of Wnt-dependent signaling in tissue regeneration and cancer. Nature Chemical Biology. 2019;**5**(2):100-107. Available from: https://www.ncbi.nlm.nih.gov/pmc/articles/PMC2628455/

[48] Sessa R, Yuen D, Wan S, Rosner M, Padmanaban P, Ge S, et al. Monocyte-derived Wnt5a regulates inflammatory lymphangiogenesis. Cell Research. 2016;**26**(2):262-265

[49] Ueno N, Wilson ME. Receptor-mediated phagocytosis of Leishmania: Implications for intracellular survival. Trends in Parasitology. 2012;**28**(8):335-344

[50] Kima PE. The amastigote forms of Leishmania are experts at exploiting host cell processes to establish infection and persist. International Journal for Parasitology. 2007;**37**(10):1087-1096

[51] Burchmore RJ, Barrett MP. Life in vacuoles—Nutrient acquisition by Leishmania amastigotes. International Journal for Parasitology. 2001;**31**(12):1311-1320

[52] Verma JK, Rastogi R, Mukhopadhyay A. Leishmania donovani resides in modified early endosomes by upregulating Rab5a expression via the downregulation of miR-494. PLoS Pathogens. 2017;**13**(6):e1006459

[53] Real F, Mortara RA. The diverse and dynamic nature of *Leishmania parasitophorous* vacuoles studied by multidimensional imaging. PLOS Neglected Tropical Diseases. 2012;**6**(2):e1518

[54] Contreras I, Gómez MA, Nguyen O, Shio MT, McMaster RW, Olivier M. Leishmania-induced inactivation of the macrophage transcription factor AP-1 is mediated by the parasite metalloprotease GP63. PLoS Pathogens. 2010;**6**(10):e1001148

[55] Lima MH, Sacramento LA, Quirino GF, Ferreira MD, Benevides L, Santana AKM, et al. Leishmania infantum parasites subvert the host inflammatory response through the adenosine A2A receptor to promote the establishment of infection. Frontiers in Immunology. 2017;**8**:815

[56] Srivastav S, Kar S, Chande AG, Mukhopadhyaya R, Das PK. Leishmania donovani exploits host deubiquitinating enzyme A20, a negative regulator of TLR signaling, to subvert host immune response. Journal of Immunology (Baltimore, MD 1950). 2012;**189**(2):924-934

[57] Novais FO, Nguyen BT, Beiting DP, Carvalho LP, Glennie ND, Passos S, et al. Human classical monocytes control the intracellular stage of Leishmania braziliensis by reactive oxygen species. The Journal of Infectious Diseases. 2014;**209**(8):1288-1296

[58] Murray HW, Rubin BY, Rothermel CD. Killing of intracellular Leishmania donovani by lymphokine-stimulated human mononuclear phagocytes. Evidence that interferon-gamma is the activating lymphokine. The Journal of Clinical Investigation. 1983;**72**(4):1506-1510

[59] Luo T, Dunphy PS, Lina TT, McBride JW. Ehrlichia chaffeensis exploits canonical and noncanonical host wnt signaling pathways to stimulate phagocytosis and promote intracellular survival. Infection and Immunity. 2015;**84**(3):686-700

[60] Abidin BM, Hammami A, Stäger S, Heinonen KM. Infection-adapted

emergency hematopoiesis promotes visceral leishmaniasis. PLoS Pathogens. 2017;**13**(8):e1006422

[61] Corrigan PM, Dobbin E, Freeburn RW, Wheadon H. Patterns of Wnt/Fzd/LRP gene expression during embryonic hematopoiesis. Stem Cells and Development. 2009;**18**(5):759-772

[62] Chae W-J, Ehrlich AK, Chan PY, Teixeira AM, Henegariu O, Hao L, et al. The Wnt antagonist Dickkopf-1 promotes pathological type 2 cell-mediated inflammation. Immunity. 2016;**44**(2):246-258

Chapter 9

Biological Role of Chalcones in Medicinal Chemistry

Sunil Tekale, Samson Mashele, Ofentse Pooe, Shivaji Thore, Pravin Kendrekar and Rajandra Pawar

Abstract

Chalcones are promising synthons and bioactive scaffolds of great medicinal interest due to their numerous pharmacological and biological activities. They are well recognized to possess antimicrobial, anticancer, antitubercular, antioxidant, anti-inflammatory, antileishmanial, and other significant biological activities. This chapter highlights recent updates and applications of chalcones as biologically, pharmacologically, and medicinally important entities.

Keywords: synthesis, characterization, PK/PD study

1. Introduction

Chalcones (trans-1,3-diaryl-2-propen-1-ones) (**1**) are α,β-unsaturated ketones consisting of two aromatic rings (ring A and B) having diverse array of substituents (**Figure 1**). Chalcone skeleton contains two aromatic rings linked by an aliphatic three-carbon chain. The two rings of chalcone are interconnected by a highly electrophilic three-carbon α,β-unsaturated carbonyl system that assumes linear or nearly planar structure. They possess conjugated double bonds and a completely delocalized π-electron system on both the aromatic rings.

Chalcones, named so by Kostanecki and Tambor, are commonly known by different names such as benzylideneacetophenone, phenyl styryl ketone, *β*-phenylacrylophenone *α*-phenyl-*β*-benzoylethylene, etc. and constitute the central core of biologically active heterocyclic compounds. Chalcones constitute good synthons for a variety of novel heterocycles of high therapeutic potential and good pharmaceutical profile [1, 2]. Chalcones themselves are identified as interesting entities associated with several biological activities [3].

The structural modifications of the chalcone rings have led to a high degree of diversity that has proven useful for the development of new medicinal agents, and thus chalcones have become an object of continued interest in both academia and industry. The chalcones are well documented for a broad spectrum of biological activities including antimicrobial, anticancer, cytotoxic, antioxidative, anti-inflammatory, antiviral, and others [4]. Currently, chalcone derivatives have been widely used for the treatment of viral disorders, cardiovascular diseases, stomach cancer, food additives, and cosmetic formulation ingredients [5]. However, much of the pharmacological potential of chalcones and their recent updates need to be understood. The purpose of this chapter is to cover and describe the recent developments, preferably after 2015 to date, the utility of chalcones as medicinally

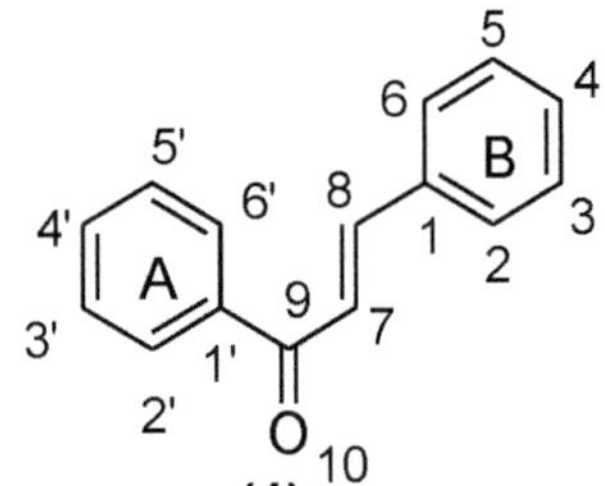

(1)

Figure 1.
Structure of chalcone.

significant scaffolds, and their biological activities. It covers and highlights the recent advances in the use of chalcones as antimicrobial, anticancer, antitubercular, antioxidant, anti-inflammatory, and miscellaneous applications in biological and medicinal fields.

2. Biological activities of chalcones

2.1 Antimicrobial chalcones

Antimicrobial agents are the drugs used to treat infectious diseases caused by different types of bacteria and fungi. The use of these drugs is now common, and continuous efforts are put by the scientific community to search for newer antimicrobial agents due to antimicrobial resistance (AMR) shown by the microbes. Mutation, gene transfer, phenotypic change, and selective pressure are some of the causes behind AMR [6]. Antimicrobial or drug resistance is commonly developed by bacteria, fungi, parasites, and viruses when the microbe no longer responds to a drug that previously treated them effectively. This AMR can lead to several issues including difficulty in controlling the disease, a longer stay of the microbes in the host, higher risks of spreading, and increase in mortality rates. Infectious diseases are one of the common problems encountered globally. Although several commercially marketed drugs are available, the search for new drug molecules becomes essential for the treatment of infectious diseases [7]. Consequently, the search for new antimicrobial agents becomes essential. Herein we discuss the recent updates in the search of chalcones as an attempt to develop antimicrobial agents:

Methoxy-4'-amino chalcones (**2**) showed good in vitro antimicrobial activities against *Escherichia coli*, *Staphylococcus aureus*, and *Candida albicans*. A molecular docking study also supported the observed results showing good interactions with the active sites of dihydropteroate synthase enzyme of *E. coli* and *S. aureus* [8].

The quinoxalinyl chalcones (**3**) synthesized by the Claisen-Schmidt condensation were found to be good antimicrobial agents. The antimicrobial studies were carried out against *Staphylococcus aureus*, *Escherichia coli*, and *Candida albicans* using the disk diffusion method. The selected chalcones were evaluated for anticancer and cytotoxicity activity against MCF-7 cancer cell lines using the MTT assay method showing good anticancer activity [9].

Some fluorinated chalcone-triazole hybrids (**4**) were studied for antimicrobial activities against *S. epidermidis*, *B. subtilis*, *E. coli*, and *P. aeruginosa* bacterial and two fungal strains, namely, *A. niger* and *C. albicans*, by standard serial dilution method [10]. The results of the in vitro antimicrobial activity were compared with ciprofloxacin and fluconazole standard drugs.

Dehydroacetic acid chalcone-1,2,3-triazole hybrids **(5)** were shown to possess good in vitro antimicrobial activities against *Staphylococcus epidermidis*, *Bacillus subtilis*, *Escherichia coli*, and *Pseudomonas aeruginosa* bacteria and two fungal strains, viz., *Aspergillus niger* and *Candida albicans* [11].

Thiazole-based chalcones including thiazolo[2,3-*b*]quinazoline and pyrido[4,3-*d*] thiazolo[3,2-*a*]pyrimidine analogs **(6** and **7)** screened against both gram-positive and gram-negative bacteria revealed that the tilted compounds had minimum inhibitory concentration (MIC) values in the ranges of 1–4.0 μg/ml against *S. aureus*, *B. subtilis*, *M. luteus*, *E. coli*, and *P. aeruginosa* [12]. The results were found to be comparable with the ampicillin and ciprofloxacin standards.

Burmaoglu et al. reported antimicrobial activity of fluoro-substituted chalcones **(8)** and **(9)** against *S. aureus*, *S. pyogenes*, *E. faecalis*, *E. coli*, and *P. aeruginosa* bacteria and *C. albicans*, *C. glabrata*, and *C. parapsilosis* fungal strains. Some of the tested compounds also exhibited antitubercular activity against *Mycobacterium tuberculosis* [13].

Chalcones incorporated with a piperazine ring **(10)** exhibited promising antimicrobial activity against *Escherichia coli*, *Aspergillus niger*, *Salmonella typhi*, *Penicillium chrysogenum*, and *Staphylococcus aureus* bacterial strains as well as *Aspergillus flavus*, *Bacillus subtilis*, and *Candida albicans* fungi [14].

Talniya and Sood documented the synthesis and antibacterial activity of chalcones **(11)** against *Bacillus subtilis* bacteria and *Aspergillus niger* fungi by disk diffusion method [15]. The chalcones possessing *o*-chloro, *p*-chloro, and *p*-hydroxyl substituents showed remarkable antimicrobial activity against the screened microbes.

Oxazolidinones incorporated with chalcone hybrids **(12)** were evaluated for in vitro antibacterial and antifungal activities by using the serial dilution method [16]. Results showed moderate antimicrobial activities as compared with the standard drugs ciprofloxacin and linezolid.

Novel diarylsulfonylurea-chalcone hybrids **(13)** were evaluated by agar well diffusion method against various strains of bacteria and fungi including *Bacillus subtilis*, *Escherichia coli*, *Bacillus pumilus*, *Staphylococcus aureus*, *Micrococcus luteus*, *Candida albicans*, and *Penicillium chrysogenum.* Most of the compounds showed promising antibacterial and antifungal activity suggesting that the diarylsulfonylurea-chalcone hybrids can be used for the treatment of diseases caused by these microbial organisms [17].

Vanillin moiety containing chalcones **(14)**, **(15)**, and **(16)** were synthesized by the Claisen-Schmidt condensation of vanillin with different acetophenone derivatives and were studied for antimicrobial activities by using agar disk diffusion and microdilution methods [18]. The researchers found *S. aureus* and *C. albicans* to be the most sensitive strains and *E. faecalis* to be the least sensitive against these chalcones. The presence of halogens in chalcones increased their microbial susceptibility. The structures of some antimicrobial chalcones are shown in **Figure 2**.

2.2 Anticancer chalcones

Cancer is a widely spreading disease all over the world, necessitating the need to develop new anticancer agents [19]. Anticancer or antineoplastic drugs are those that are effective in the treatment of malignant or cancerous disease. Increasing recurrence of mammalian tumors and severe side effects of chemotherapeutic agents reduce the clinical efficiency of a large variety of commonly used anticancer agents, and thus, there is always a constant need to develop alternative or synergistic anticancer drugs with minimal side effects [20].

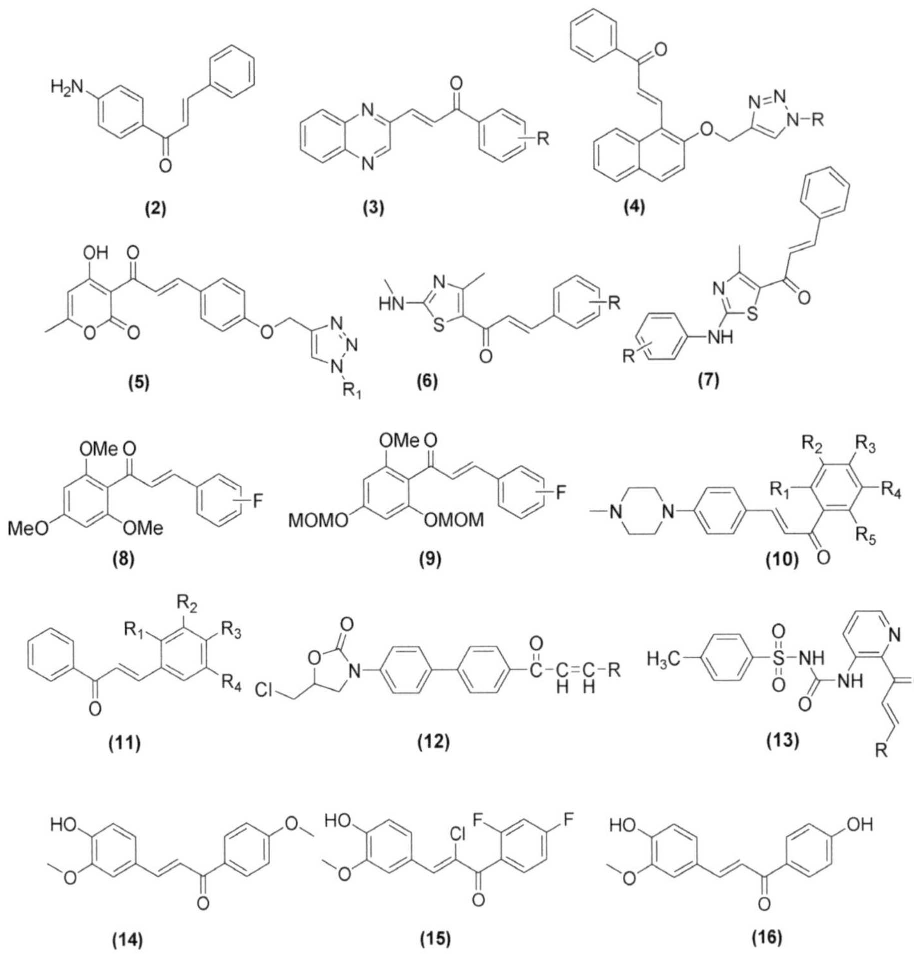

Figure 2.
Structures of antimicrobial chalcones.

The treatment of cancer is a complicated process as the drugs used target human cells and albeit cells that have undergone genetic changes and are dividing at a fast and uncontrolled rate. However, only a few anticancer drugs can differentiate between normal tissue cells and cancer cells to a large extent. Thus, there is always a constant need to develop alternative or synergistic anticancer drugs with minimal side effects. This part of the present chapter highlights significant and recent developments in chalcones used as anticancer agents:

Sulfonylpiperazines linked with [1,3]dioxolo[4,5-*g*]chromenones (17) were synthesized by the aldol condensation and evaluated as antioxidants against DPPH, ABTS, as well as antiproliferative agents against non-cancer MDCK cell lines [21].

The design, synthesis, and antitumor potential of chalcones **(18)** were studied against human breast adenocarcinoma MCF-7 cells in a concentration-dependent manner [22]. They triggered significant changes in cell morphology and biochemical/molecular parameters and revealed the apoptosis inductor nature of the titled compounds and their application as promising alternatives for the treatment of neoplasia, especially in terms of drug resistance development.

Novel anthraquinone-chalcone hybrids **(19)** possessing amide functionality were synthesized, then characterized, and reported for good cytotoxic potential against K562, Jurkat, and HL-60 leukemia cell lines [23].

An apoptosis is an important phenomenon, which affects many diseases, such as cancer and Alzheimer's disease. Chalcones **(20)** induced apoptosis of human hepatic and lung cancer cells and inhibited cancer cell migration and invasion [24].

The bis-chalcone derivatives **(21)** were studied for their ability to inhibit xanthine oxidase and growth inhibitory activity against MCF-7 and caco-2 human cancer cell lines in vitro. The bis-chalcone with fluoro group at the 2nd or 2, 5th position of B-ring was found to be a potent inhibitor of the enzyme possessing IC_{50} values in the low micromolar range. The activities of the compounds were found to be around seven times higher than the standard allopurinol [25].

Chalcones **(22)** were synthesized and evaluated for anticancer activities on human colorectal carcinoma cell line HCT116 by Dias et al. [26]. Halogens at the third position of the chalcones were found to enhance the anticancer activity of the titled compounds.

Leao et al. reported the chalcone derivatives **(23)** and **(24)** for cytotoxicity against human tumor cells [27]. Some novel xanthine-chalcone hybrids **(25)** and **(26)** were reported as promising anticancer agents [28].

A series of novel dithiocarbamate-chalcone derivatives **(27)** and **(28)** was designed, synthesized, and evaluated for antiproliferative activity against three selected cancer cell lines (EC-109, SK-N-SH, and MGC-803). Almost all the synthesized compounds exhibited moderate to potent activity against all the tested cancer cell lines [29].

Pd(II) and Pt(II) complexes of chalcones **(29)** were studied for in vitro antimicrobial and antitumor activities against different microorganisms and the human hepatocellular carcinoma cells indicating their use as promising antimicrobial agents and anticancer drug candidates [30].

Some novel Pt(IV) complexes of chalcone analogs **(30)** were synthesized and evaluated for antiproliferative activity by using MTT assay. The in vitro evaluation revealed that all Pt(IV) complexes showed good activity against the three human cancer cells [31].

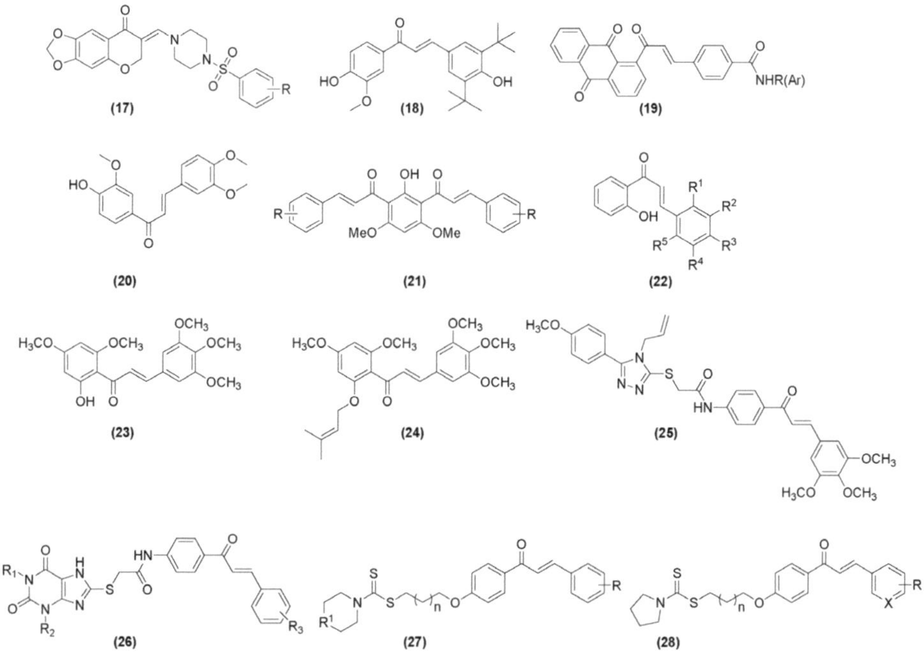

Figure 3.
Structures of anti-cancer chalcones (17-28).

The overexpression of the CYP1 class of enzymes is associated with the development of human carcinomas. The pyridine-4-yl series of chalcones **(31)** were synthesized and screened for the inhibition of CYP1 isoforms in Sacchrosomes TM and live human HEK293 cells. The chalcones bearing tri-alkoxy groups on non-heterocyclic ring displayed selective inhibition of the CYP1A1 enzyme with IC_{50} values less than 70 nM [32].

The pyrazolic chalcone analogous compounds **(32)** were synthesized and evaluated as potential chemotherapeutic agents for the treatment of hepatocellular carcinoma [33]. Some of the screened compounds exhibited potent cytotoxic activity against all the cancer cell lines tested and had good cytotoxic activities.

DNA ligases play a crucial role in causing cancer. Gupta et al. reported the inhibition DNA ligases resulting in DNA nick-sealing activity followed by the antiproliferative activity of the indole-chalcone based benzopyran chalcones **(33)** on cancer cells [34].

Indolizine-chalcone hybrids **(34)** were synthesized and studied for apoptosis and anticancer effect on human lymphoma cells by S. Park and coworkers [35].

Prenyl and geranyl group-bearing chalcones **(35)** were synthesized by using regioselective iodination followed by the Suzuki coupling reaction and studied for in vitro anticancer activity against human tumor cell line K562 by MTT assay. Morphology changes revealed that the chalcone derivatives inhibited the proliferation of K562 cells by inducing apoptosis [36].

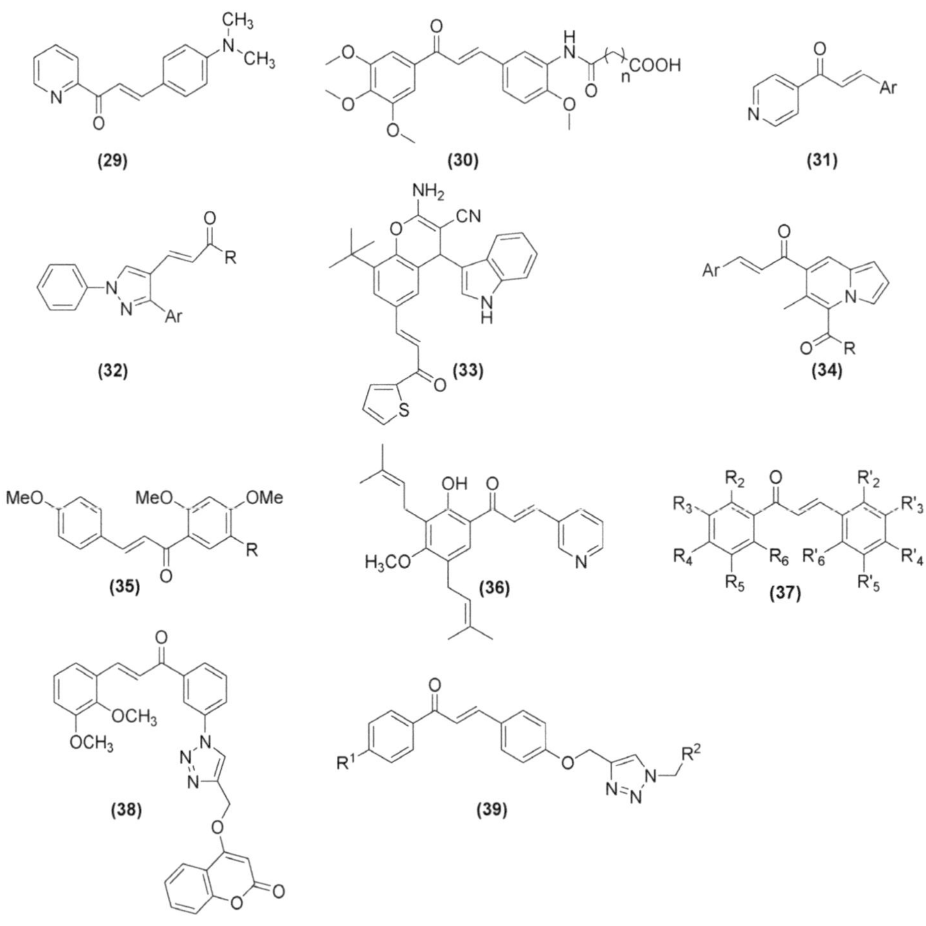

Figure 4.
Structures of anticancer chalcones (29-39).

Leukemia is a hematologic malignancy with poor prognosis in humans. Diprenylated chalcone **(36)** was studied as a new potential antileukemia agent [37].

Chalcones **(37)** were studied for antiproliferative activities against the human TRAIL-resistant breast (MCF-7, MDA-MB-231), cervical (HeLa), ovarian (Caov-3), lung (A549), liver (HepG2), colorectal (HT-29), nasopharyngeal (CNE-1), erythromyeloblastoid (K-562), and T-lymphoblastoid (CEM-SS) cancer cells by Mai [38].

Triazole incorporated with chalcones **(38)** and **(39)** was synthesized and evaluated for 3-(4,5-dimethylthiazole-2-yl)-2,5-diphenyltetrazolium bromide (MTT) cytotoxicity assay against a series of four human cancer cell lines (MCF-7, MIA-PaCa-2, A549, HepG2) [39]. The structures of anticancer chalcones are depicted in **Figures 3** and **4**.

2.3 Antitubercular chalcones

Tuberculosis (TB), caused by the acid-fast gram-positive bacillus, *Mycobacterium tuberculosis*, remains the leading source of bacterial infectious disease [40]. *M. tuberculosis* establishes an infection through an invasion of alveolar macrophages. The *Mycobacterium tuberculosis* encodes for more than 60 adenylating enzymes, mainly tRNA synthetases, acyl-AMP ligases, etc. [41]. Currently, the treatment of TB employs four first-line drugs, isoniazid, rifampin, pyrazinamide, and ethambutol, which must be administered in the body daily for a 2-month intensive phase. However, for susceptible TB strains, this therapy is 95% effective. The emergence of multidrug-resistant (MDR) strains, defined as resistant to isoniazid and rifampin, requires the use of less effective and more toxic second-line TB drugs. Herein we discuss some recent updates in the application of chalcones against tuberculosis:

New sulfonamide-bearing chalcones **(40)** were synthesized by the Claisen-Schmidt condensation and were reported as excellent antituberculosis hits showing low selectivity, being equally inhibitory to *M. tuberculosis* and mammalian T3T cells [42].

Gomes et al. studied antitubercular activities of chalcones **(41)** and **(42)**. The chalcones showed good selectivity towards *M. tuberculosis* with low cytotoxicity against Vero cells and thus possess promising antitubercular potential [43].

Spirochromone annulated chalcone conjugates (43) were documented for antitubercular activity against *Mycobacterium tuberculosis* H37Rv strain. Molecular docking studies performed against the receptors revealed MTB phosphotyrosine phosphatase B protein as the most probable target based on the high binding-affinity scores [44].

Babu et al. studied chalcones containing nitrophenyl moieties **(44)** for antitubercular activity using MABA assay and antibacterial and antifungal activities by cup plate method. Molecular docking study predicted the inhibition of thymidine kinase of the *Mycobacterium tuberculosis* [45]. Anti-tubercular chalcones are depicted in **Figure 5**.

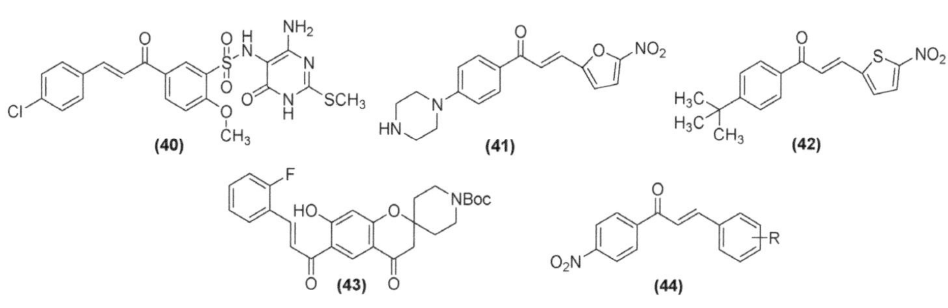

Figure 5.
Structures of anti-tubercular chalcones.

2.4 Antioxidant chalcones

Antioxidants are the compounds that inhibit the oxidation process. These substances can prevent or slow damage to cells caused by free radicals. Oxidation is a chemical reaction that generates free radicals, thereby leading to chain reactions which may damage the cells of organisms and hence responsible for oxidative stress resulting in chronic diseases such as heart diseases, stroke, cancer, arthritis, respiratory diseases, Parkinson's disease, and other inflammatory conditions [46].

Cao et al. documented a series of 4′-OH-flurbiprofen-chalcone hybrids **(45)** and evaluated them as potential multifunctional agents for the treatment of Alzheimer's disease. Besides, the compounds were reported for good antioxidant activities, MAO inhibitions, biometal chelating abilities, and in vitro anti-neuroinflammatory activities [47].

Selenoenzymes and nuclear factor erythroid 2-related factor 2 (Nrf2)-regulated phase II enzymes constitute the main components of cellular redox and antioxidant systems giving information about multiple interrelations involved in the oxidation processes. Chalcones **(46)** were proved to interfere with the biosynthesis of Nrf2-regulated selenoenzymes [48].

El-Sayed et al. documented the antioxidant activity of chalcones **(47)** [49].

The chalcone derivatives **(48)** were synthesized by the Claisen-Schmidt condensation with KOH in ethanol at room temperature under sonication conditions and screened for antioxidant potential by Polo et al. [50].

The chalcones **(49)** were studied as potent antioxidants by Tajammal and coworkers. These compounds have lower IC_{50} values than the Trolox and ascorbic acid standards [51].

A series of chalcone **(50)** analogs were designed, synthesized, and screened for antioxidant activities. The chalcone was found as a promising anti-ischemic stroke drug candidate, providing novel dual-antioxidant mechanism strategies and concepts for oxidative stress-related disease treatment [52]. The prenylated chalcones **(51)** were reported for good antioxidant activity [53]. **Figure 6** represents the structures of antioxidant chalcones.

2.5 Anti-inflammatory chalcones

Anti-inflammatory drugs are the drugs which are used to reduce pain and inflammation. In other words, these are pain-relieving drugs. These drugs work mainly by inhibiting the cyclooxygenase enzymes, COX-1 and COX-2, that produce

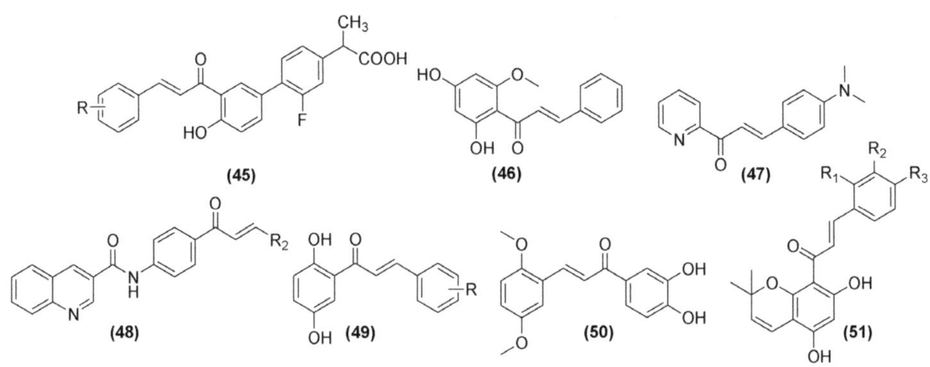

Figure 6.
Structures of antioxidant chalcones.

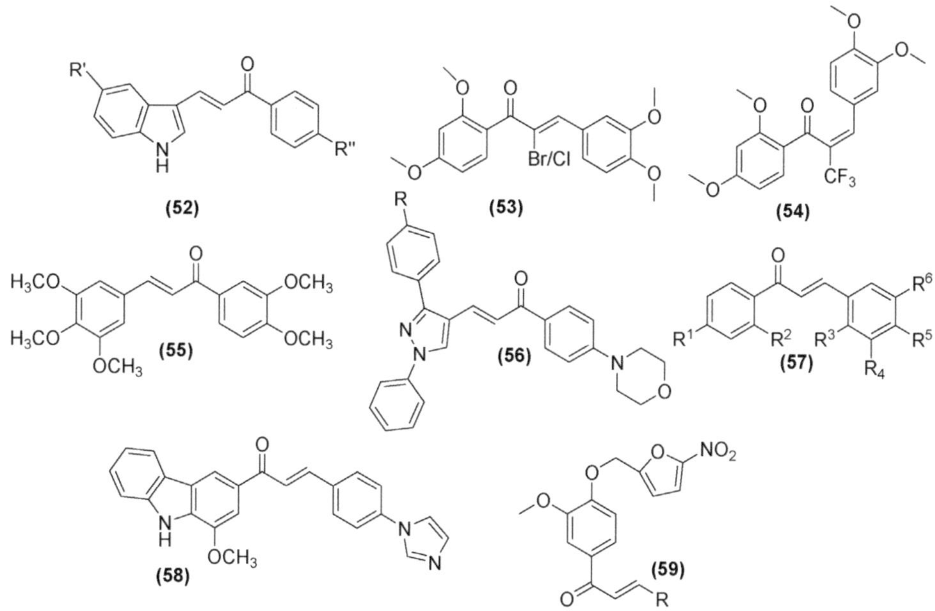

Figure 7.
Structures of anti-inflammatory chalcones.

prostaglandins [54]. Herein we discuss some of the efforts for the development of chalcone-based heterocycles as effective anti-inflammatory compounds:

Indole-based chalcones **(52)** were synthesized and evaluated for in vitro COX-1 and COX-2 inhibitory activity [55].

α-Substituted 2′,3,4,4′-tetramethoxychalcones (53) and (54) were evaluated for their ability to modulate inflammatory responses to influence on heme oxygenase-1, nitric oxide synthase, and cytokine expression levels. Anti-inflammatory activity was correlated with thiol-alkylating activity, i.e., stronger electrophiles substituted with CF_3, Br, and Cl were found to be more potent than the remaining derivatives [56].

Zhang et al. identified methoxy chalcones **(55)** as a potential candidate for treating acute inflammatory diseases [57].

Pyrazole- and morpholine-containing chalcones **(56)** were reported for anti-inflammatory activity by Gadhave and Uphade. The anti-inflammatory activity performed by carrageenan-induced rat paw edema method showed good potency of some of the tested compounds as compared with the standard diclofenac drug [58].

Nurkenov et al. studied the in vitro anti-inflammatory effect of chalcones **(57)** to inhibit the lipopolysaccharide-induced production of anti-inflammatory cytokine interleukin-6 and tumor necrosis factor [59].

The imidazole containing chalcone molecule **(58)** demonstrated noteworthy anti-inflammatory activity as compared with the standard drug, indomethacin [60].

1-[3-Methoxy-4-(5-nitro-furan-2-ylmethoxy)-phenyl]-3-(substituted phenyl)-propenones **(59)** synthesized by the condensation of furfural and apocynin were evaluated for anti-inflammatory activity [61]. The structures of anti-inflammatory chalcones are shown in **Figure 7**.

2.6 Miscellaneous applications of chalcones

Besides the above-discussed applications, chalcones are useful for miscellaneous applications. Some of them are mentioned as follows:

Leishmania is a genus of trypanosomes responsible for the disease leishmaniasis. Leishmaniasis is spread through sand flies of the genus *Phlebotomus*, primary hosts being the vertebrates. The chalcone **(60)** was evaluated against 29 promastigotes of *Leishmania donovani* exhibiting low toxicity against mammalian cells [62].

A series of new chalcone-rivastigmine hybrids (61) was designed, synthesized, and evaluated **in vitro** for the ability to inhibit human acetylcholinesterase and butyrylcholinesterase. Results showed that these compounds exhibited selective activity in micro- and submicromolar ranges as compared with the standard rivastigmine and thus the compounds can serve as the lead ones for the treatment of Alzheimer's disease [63].

Sang et al. reported AChE/BChE inhibitory, MAO-A/MAO-B inhibitory, and antioxidant activities of chalcone-O-carbamate derivatives **(62)**. Results revealed that the compounds show highly selective BChE inhibitory activity with IC50 values of 1–3 mM range [64].

Some 1,3,4-oxadiazole/thiadiazole-chalcone conjugates **(63)** were synthesized and evaluated for in vitro and in vivo antiviral activities against TMV. These conjugates have low binding constant values which were comparable to the standard ningnanmycin [65].

Amide tethered 7-chloroquinoline-chalcone bifunctional hybrids (64) were synthesized and employed as antimalarial agents against the resistant strain of *Plasmodium falciparum*. Methoxy substituent at the *para* position of ring B on chalcones and longer alkyl chain lengths significantly improved the antiplasmodial profiles of the chalcone derivatives [66].

The halogenated 1-tetralone or 6-amino-1-tetralone chalcone derivatives **(65)** were synthesized and evaluated for inhibitory effects against ROS production in LPS-stimulated RAW 264.7 macrophages. The structure-activity relationship revealed that amino moiety at the sixth position of 1-tetralone chalcones plays an important role for greater ROS inhibitory potency [67].

Chalcone derivatives (**66**) were studied for hepatoprotective ability, and the results were compared with the standard hepatoprotective drug silymarin. The experimental results were supported by a molecular docking study [68].

Triazole-linked 4-aminoquinoline-chalcone/-*N*-acetylpyrazoline conjugates **(67)** were synthesized and evaluated for antiplasmodial activities against cultured chloroquine-resistant strain. The activities were found to be dependent on the length of the alkyl chain as well as on the presence of methoxy substituents on the chalcone rings [69].

Chalcone analogs **(68)** were synthesized and evaluated for cytotoxic effects in human hepatoma HepG2 cells. The percentage of apoptotic cells was significantly higher in the compounds than that in the control cells [70].

The oxygenated chalcones **(69)** were found to inhibit monoamine oxidases, and the lead compounds were found to be nontoxic at 200 μg/mL in normal rat spleen cells [71].

Hameed et al. studied the quinoline-based chalcone compounds **(70)** as reverse transcriptase inhibitors. Bromo- and chloro-substituted chalcones exhibited a high degree of inhibition against the reverse transcriptase [72].

Histoplasmosis is a fungal infection caused by the dimorphic fungus *Histoplasma* 27 *capsulatum*. Hydroxyl group-bearing chalcones **(71)** and **(72)** were studied for histoplasmosis by Wanessa et al. [73].

Sashidhara et al. documented the antiulcer activity of some novel quinoline-chalcone hybrids **(73)** in various ulcer models in Sprague Dawley rats. Additional studies including in vitro metabolic stability and in vivo pharmacokinetics showed their potential to act as an orally active and safe candidate for the development of an antiulcer agent [74].

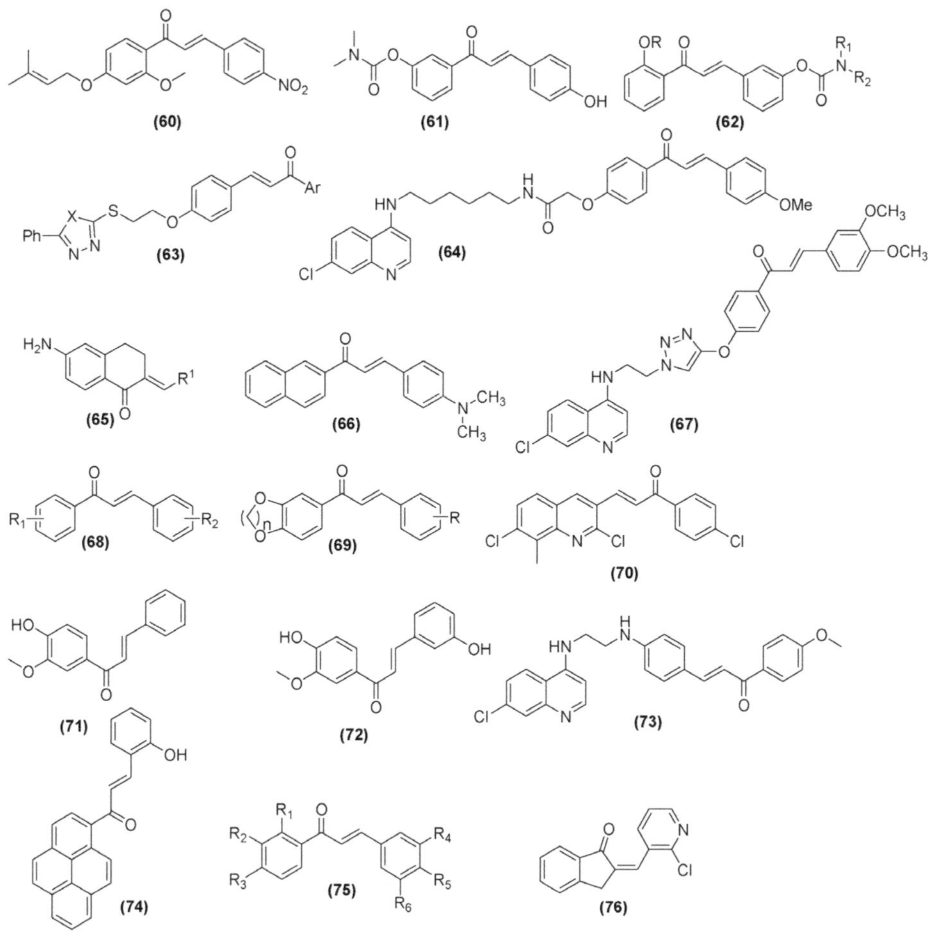

Figure 8.
Structures of chalcones having miscellaneous activities.

Pyrene ring-bearing chalcone **(74)** was studied as a sensitive and highly selective sensor for the detection of aluminum (Al^{3+}) ions by fluorimetric studies by Suresh et al. The chalcone was found to be useful for the electrosorptive removal of Al^{3+} ion and several other biological applications including the bio-imaging of bacterial cells [75].

Chalcones **(75)** were reported for potent antimalarial activities against *Plasmodium falciparum* using Rieckmann's method. Allyloxy, hydroxy, and alkoxy functional groups increased the antimalarial activity of the chalcone derivatives [76].

Human African trypanosomiasis is an infectious disease that affects the lives of people living in rural areas of Africa. Beteck et al. studied the antitrypanosomal activities of indanone-based chalcone analogs **(76)** by screening against *T.b. brucei* [77]. The structures of chalcones having miscellaneous activities are depicted in **Figure 8**.

3. Conclusion

Chalcones and their analogs possess significant biological activities including antimicrobial, anticancer, antitubercular, antioxidant, anti-inflammatory, antileishmanial, enzyme inhibitory, and miscellaneous applications and hence acquire a unique place in medicinal chemistry. The growing interest of synthetic organic, pharmacological, and medicinal chemists towards chalcones and their derivatives will be continued in the future also. This chapter is expected to provide a stimulus for researchers to design, synthesize, and carry out further investigation on the pharmacological effects of new chalcone derivatives for different biological activities.

Conflict of interest

The authors declare no conflict of interest.

Author details

Sunil Tekale[1], Samson Mashele[2], Ofentse Pooe[3], Shivaji Thore[1], Pravin Kendrekar[2*] and Rajandra Pawar[1*]

1 Department of Chemistry, Deogiri College, Aurangabad, MS, India

2 Faculty of Health and Environmental Sciences, Department of Health Sciences, Free State, Bloemfontein, South Africa

3 Discipline of Biochemistry, School of Life Science, University of KwaZulu-Natal, Durban, South Africa

*Address all correspondence to: kkpravin@gmail.com and rppawar@yahoo.com

IntechOpen

© 2020 The Author(s). Licensee IntechOpen. This chapter is distributed under the terms of the Creative Commons Attribution License (http://creativecommons.org/licenses/by/3.0), which permits unrestricted use, distribution, and reproduction in any medium, provided the original work is properly cited.

References

[1] Zhuang C, Zhang W, Sheng C, Zhang W, Xing C, Miao Z. Chalcone: A privileged structure in medicinal chemistry. Chemical Reviews. 2017; **117**(12):7762-7810

[2] Gomes MN, Muratov EN, Pereira M, Peixoto JC, Rosseto LP, Cravo PVL, et al. 1,3 chalcone derivatives: Promising starting points for drug design. Molecules. 2017;**22**(1210):1-25

[3] Verma S, Srivastava AK, Pandey OP. A review on chalcones synthesis and their biological activity. Pharma Tutor. 2018;**6**(2):22-39

[4] Chavan BB, Gadekar AS, Mehta PP, Vawhal PK, Kolsure AK, Chabukswar AR. Synthesis and medicinal significance of chalcones—A review. Asian Journal of Biomedical and Pharmaceutical Sciences. 2016;**6**(56): 01-07

[5] Jaiswal P, Pathak DP, Bansal H, Agarwal U. Chalcone and their heterocyclic analog: A review article. Journal of Chemical and Pharmaceutical Research. 2018;**10**(4):160-173

[6] Livermore DM. The need for new antibiotics. Clinical Microbiology and Infection. 2004;**10**(4):1-9

[7] Jackson N, Czaplewski L, Piddock LJV. Discovery and development of new antibacterial drugs: Learning from experience? Journal of Antimicrobial Chemotherapy. 2018; **73**(6):1452-1459

[8] Suwito H, Matuzahroh N, Kristanti AN, Hayati S, Dewi SR, Amalina I, et al. Antimicrobial activities and in silico analysis of methoxy amino chalcone derivatives. Procedia Chemistry. 2016;**18**:103-111

[9] Desai V, Desai S, Gaonkar SN, Palyekar U, Joshi SD, Dixit SK. Novel quinoxalinyl chalcone hybrid scaffolds as enoyl ACP reductase inhibitors: Synthesis, molecular docking, and biological evaluation. Bioorganic & Medicinal Chemistry Letters. 2017;**27**:2174-2180

[10] Yadav P, Lal K, Kumar L, Kumar A, Kumar A, Paul AK, et al. Synthesis, crystal structure and antimicrobial potential of some fluorinated chalcone-1,2,3-triazole conjugates. European Journal of Medicinal Chemistry. 2018; **155**:263-274

[11] Lal K, Yadav P, Kumar A, Kumar A, Paul AK. Design, synthesis, characterization, antimicrobial evaluation and molecular modeling studies of some dehydroacetic acid-chalcone-1,2,3-triazole hybrids. Bioorganic Chemistry. 2018;**77**:236-244

[12] Alrohily WD, Habib ME, El-Messery SM, Alqurshi A, El-Subbagh H, Habib ESE. Antibacterial, antibiofilm and molecular modeling study of some antitumor thiazole based chalcones as a new class of DHFR inhibitors. Microbial Pathogenesis. 2019;**136**:103674

[13] Burmaoglu S, Algul O, Gobek A, Anil DA, Ulger M, Erturk BG, et al. Design of potent fluoro-substituted chalcones as antimicrobial agents. Journal of Enzyme Inhibition and Medicinal Chemistry. 2017:490-495

[14] Shah NN, Ziauddin HM, Zameer M, Kendre MM, Dhole JA, Baseer MA. Synthesis and antimicrobial studies of a novel series of piperazine chalcones. Der Chemica Sinica. 2011;**2**(1):33-37

[15] Noorulhaq SSN, Baseer MA, Talniya NC, Sood P. Synthesis and antimicrobial activity of chalcones. Journal of Chemical and Pharmaceutical Research. 2016;**8**(5):610-613

[16] Panigrahi N, Ganguly S, Panda J. Ultrasound-assisted synthesis,

characterization and antimicrobial evaluation of novel oxazolidinone biphenyl chalcone hybrid derivatives. Indian Journal of Pharmaceutical Education and Research. 2019;**53**(2): 286-300

[17] Naidu MA, Prasad YR. Synthesis of novel diarylsulfonylurea-chalcone hybrid molecules with potential in vitro antimicrobial activity. Asian Journal of Pharmaceutics. 2018;**12**(2): 88-93

[18] Bathelemy N, Charles FN, Pantaleon A, Azeh NN, Estella TF, Hortense GK, et al. Synthesis and evaluation of antimicrobial properties of some chalcones. British Journal of Pharmaceutical Research. 2016;**14**(2): 1-11. Article no. BJPR.28243

[19] Altman LK. A growing threat. Cancer Posit. 1997;**8**(4):21-23

[20] Xu X, Qiu G, Ji L, Ma R, Dang Z, Jia R, et al. Research and development of anticancer agents under the guidance of biomarkers. Cancer Traditional Medicine. 2019;**5**(1):17-21

[21] Patel RV, Mistry BM, Syed R, Parekh NM, Shin HS. Phenylsulfonyl piperazine bridged [1,3] dioxolo [4,5-g] chromenones as promising antiproliferative and antioxidant agents. Bioorganic Chemistry. 2019;**87**:23-30

[22] Cabral BLS, Silva ACG, Avila RI, Cortez AP, Luzin RM, Liao LM, et al. A novel chalcone derivative, LQFM064, induces breast cancer cells death via p53, p21, KIT and PDGFRA. European Journal of Pharmaceutical Sciences. 2017;**107**:1-15

[23] Stanojkovic T, Markovic V, Matic IZ, Mladenovic MP, Petrovica N, Krivokuca A, et al. Highly selective anthraquinone-chalcone hybrids as potential antileukemia agents. Bioorganic & Medicinal Chemistry Letters. 2018;**28**:2593-2598

[24] Dong N, Liu X, Zhao T, Wang L, Li H, Zhang S, et al. Apoptosis-inducing effects and growth inhibitory of a novel chalcone, in human hepatic cancer cells and lung cancer cells. Biomedicine & Pharmacotherapy. 2018;**105**:195-203

[25] Burmaoglua S, Ozcanb S, Balciogluc S, Genceld M, Nomac SAA, Essizd S, et al. Synthesis, biological evaluation and molecular docking studies of bis-chalcone derivatives as xanthine oxidase inhibitors and anticancer agents. Bioorganic Chemistry. 2019;**91**:103149

[26] Dias TA, Duarte CL, Lima CF, Proenca MF, Wilson CP. Superior anticancer activity of halogenated chalcones and flavonols over the natural flavonol quercetin. European Journal of Medicinal Chemistry. 2013;**65**:500-510

[27] Leao M, Soares J, Gomes S, Raimundo L, Ramos H, Queiroz CBG, et al. Enhanced cytotoxicity of prenylated chalcone against tumour cells via disruption of the p53-MDM2 interaction. Life Sciences. 2015;**142**: 60-65

[28] Abou-Zied HA, Youssif BGM, Mohamed MFA, Hayallah AM, Abdel-Aziz M. EGFR inhibitors, and apoptotic inducers: Design, synthesis, anticancer activity and docking studies of novel xanthine derivatives carrying chalcone moiety as hybrid molecules. Bioorganic Chemistry. 2019;**89**:102997

[29] Fu DJ, Zhang SY, Liu YC, Zhang L, Liu JJ, Song J, et al. Design, synthesis and antiproliferative activity studies of novel dithiocarbamate-chalcone derivates. Bioorganic & Medicinal Chemistry Letters. 2016;**26**(16):3918-3922

[30] Gaber M, El-Ghamry HA, Mansour MA. Pd(II) and Pt(II) chalcone complexes. Synthesis, spectral characterization, molecular modeling, biomolecular docking, antimicrobial and

antitumor activities. Journal of Photochemistry and Photobiology A: Chemistry. 2018;**354**:163-174

[31] Huang X, Huang R, Wang Z, Li L, Gou S, Liao Z, et al. Pt(IV) complexes conjugating with chalcone analog as inhibitors of microtubule polymerization exhibited selective inhibition in human cancer cells. European Journal of Medicinal Chemistry. 2018;**146**:435-450

[32] Horley NJ, Beresford KJM, Kaduskar S, Joshi P, McCann GJP, Ruparelia KC, et al. (E)-3-(3,4,5-Trimethoxyphenyl)-1-(pyridin-4-yl) prop-2-en-1-one, a heterocyclic chalcone is a potent and selective CYP1A1 inhibitor and cancer chemopreventive agent. Bioorganic & Medicinal Chemistry Letters. 2017;**27**:5409-5414

[33] Hawash MM, Kahraman DC, Eren F, Cetin Atalay R, Baytas SN. Synthesis and biological evaluation of novel pyrazolic chalcone derivatives as novel hepatocellular carcinoma therapeutics. European Journal of Medicinal Chemistry. 2017;**31**(129):12-26

[34] Gupta S, Maurya P, Upadhyay A, Kushwaha P, Krishna S, Siddiqi MI, et al. Synthesis and bio-evaluation of indole-chalcone based benzopyrans as promising antiligase and antiproliferative agents. European Journal of Medicinal Chemistry. 2018; **143**:1981-1996

[35] Park S, Kim EH, Kim J, Kim SH, Kim I. Biological evaluation of indolizine-chalcone hybrids as new anticancer agents. European Journal of Medicinal Chemistry. 2018;**144**:435-443

[36] Wang HM, Zhang L, Liu J, Yang ZL, Zhao HY, Yang Y, et al. Synthesis and anticancer activity evaluation of novel prenylated and geranylated chalcone natural products and their analogs. European Journal of Medicinal Chemistry. 2015;**92**:439-448

[37] Zhang YQ, Wen ZH, Wan K, Yuan D, Zeng X, Liang G, et al. A novel synthesized 3’,5’-diprenylated chalcone mediates the proliferation of human leukemia cells by regulating apoptosis and autophagy pathways. Biomedicine & Pharmacotherapy. 2018;**106**:794-804

[38] Mai CW, Yaeghoobi M, Abd-Rahman N, Kang YB, Pichik MR. Chalcones with electron-withdrawing and electron-donating substituents: Anticancer activity against TRAIL-resistant cancer cells, structure-activity relationship analysis and regulation of apoptotic proteins. European Journal of Medicinal Chemistry. 2014;**77**:378-387

[39] Yadav P, Lal K, Kumar A, Guru SK, Jaglan S, Bhushan S. Green synthesis and anticancer potential of chalcone linked-1,2,3-triazoles. European Journal of Medicinal Chemistry. 2017;**126**: 944-953

[40] Tiberi S, Munoz-Torrico M, Duarte R, Dalcolmo M, Ambrosio LD, Migliori GB. New drugs and perspectives for new anti-tuberculosis regimens. Pulmonology. 2018;**24**(2): 86-98

[41] Unissa AN, Hanna LE. Molecular mechanisms of action, resistance, detection to the first-line anti-tuberculosis drugs: Rifampicin and pyrazinamide in the post-whole-genome sequencing era. Tuberculosis. 2017;**105**:96-107

[42] Castano LF, Cuartas V, Bernal A, Insuasty A, Guzman J, Vidal O, et al. New chalcone-sulfonamide hybrids exhibiting anticancer and antituberculosis activity. European Journal of Medicinal Chemistry. 2019; **176**:50-60

[43] Gomes MN, Braga RC, Grzelak EM, Neves BJ, Muratov E, Ma R, et al. QSAR-driven design, synthesis and discovery of potent chalcone derivatives with antitubercular activity. European

Journal of Medicinal Chemistry. 2017; **137**:126-138

[44] Mujahid M, Yogeeswari P, Sriram D, Basavanag UMV, Cervantes ED, Córdoba-Bahena L, et al. Spirochromone-chalcone conjugates as antitubercular agents: Synthesis, bio evaluation, and molecular modeling studies. RSC Advances. 2015;**5**: 106448-106460

[45] Babu LS, Shaik AB, Prasad YR. Synthesis, antibacterial, antifungal antitubercular activities and molecular docking studies of nitrophenyl derivatives. International Journal of Life Sciences and Pharmaceutical Research. 2019;**9**(1):54-64

[46] Albrecht S, Elpelt A, Kasim C, Reble C, Mundhenk L, Pischon H, et al. Quantification and characterization of radical production in human, animal and 3D skin models during sun irradiation measured by EPR spectroscopy. Free Radical Biology and Medicine. 2019;**131**:299-308

[47] Cao Z, Yang J, Xu R, Song Q, Zhang X, Liu H, et al. Design, synthesis and evaluation of 4′-OH-flurbiprofen-chalcone hybrids as potential multifunctional agents for Alzheimer's disease treatment. Bioorganic & Medicinal Chemistry. 2018;**26**(5): 1102-1115

[48] Spirt SD, Eckers A, Wehrend C, Micoogullari M, Sies H, Stahl W, et al. Interplay between the chalcone cardamonin and selenium in the biosynthesis of Nrf2-regulated antioxidant enzymes in intestinal caco-2 cells. Free Radical Biology and Medicine. 2016;**91**:164-171

[49] El-Sayed YS, Gaber M. Studies on chalcone derivatives: Complex formation, thermal behavior, stability constant and antioxidant activity. Spectrochimica Acta Part A: Molecular and Biomolecular Spectroscopy. 2015; **137**:423-431

[50] Polo E, Ibarra-Arellano N, Prent-Penaloza L, Morales-Bayuelo A, Henao J, Galdamez A, et al. Ultrasound-assisted synthesis of novel chalcone, heterochalcone and bis-chalcone derivatives and the evaluation of their antioxidant properties and as acetylcholinesterase inhibitors. Bioorganic Chemistry. 2019;**90**:103034

[51] Tajammal A, Batool M, Ramzan A, Samra MM, Mahnoor I, Verpoort F, et al. Synthesis, antihyperglycemic activity and computational studies of antioxidant chalcones and flavanones derived from 2,5-dihydroxyacetophenone. Journal of Molecular Structure. 2017;**1148**:512-520

[52] Huang L, Cheng C, Ge L, Xie J, Shen M, et al. Design, synthesis and biological evaluation of chalcone analogues with novel dual antioxidant. Acta Pharmaceutica Sinica B. 2019;**9**(2): 335-350

[53] Teng Y, Li X, Yang K, Li X, Zhang Z, et al. Synthesis and antioxidant evaluation of desmethylxanthohumol analogs and their dimers. European Journal of Medicinal Chemistry. 2017; **125**:335-345

[54] Vane JR, Botting RM. Anti-inflammatory drugs and their mechanism of action. Inflammatory Research. 1998;**47**(Suppl 2):S78-S87

[55] Ozdemir A, Altıntop MD, Turan-Zitouni G, Çiftçi GA, Ertorun I, Alataş O, et al. Synthesis and evaluation of new indole-based chalcones as potential antiinflammatory agents. European Journal of Medicinal Chemistry. 2015; **89**:304-309

[56] Rucker H, Al-Rifai N, Rascle A, Gottfried E, Brodziak-Jarosz L, Gerhauser C, et al. Enhancing the anti-inflammatory activity of chalcones by tuning the Michael acceptor site.

Organic and Biomolecular Chemistry. **2015**;**13**:3040-3047

[57] Zhang Y, Wu J, Ying S, Chen G, Wu B, Xu T, et al. Discovery of new MD2 inhibitor from chalcone derivatives with anti-inflammatory effects in LPS-induced acute lung injury. Scientific Reports. 2016;**6**. Article number: 25130

[58] Gadhave AG, Uphade BK. Synthesis of some pyrazole containing chalcones and pyridine-3-carbonitriles and study of their anti-inflammatory activity. Oriental Journal of Chemistry. 2017; **33**(1):219-225

[59] Nurkenov OA, Ibraev MK, Schepetkin IA, Khlebnikov AI, Seilkhanov TM, Arinova AE, et al. Synthesis, structure, and anti-inflammatory activity of functionally substituted chalcones and their derivatives. Russian Journal of General Chemistry. 2019;**89**(7): 1360-1367

[60] Mahapatra DK, Shivhare RS. Anti-inflammatory potential of a novel imidazole containing murrayanine based chalcone. Modern Applications in C Pharmacy & Pharmacology. 2018; 2(2):1-4. MAPP.000533

[61] Kumar Reddy ALV, Kathale NE. Synthesis, characterization and anti-inflammatory activity of chalcone derivatives linked with apocynin and 5-nitrofuran moiety. Asian Journal of Chemistry. 2018;**30**(2):312-316

[62] Ortalli M, Ilari A, Colotti G, Ionna ID, Battista T, Bisi A, et al. Identification of chalcone-based antileishmanial agents targeting trypanothione reductase. European Journal of Medicinal Chemistry. 2018;**152**:527-541

[63] Wang L, Wang Y, Tian Y, Shang J, Sun X, Chen H, et al. Design, synthesis, biological evaluation, and molecular modeling studies of chalcone-rivastigmine hybrids as cholinesterase inhibitors. Bioorganic & Medicinal Chemistry. 2017;**25**(1):360-371

[64] Sang Z, Wang K, Shi J, Liu W, Tan Z. Design, synthesis, in-silico and biological evaluation of novel chalcone-O-carbamate derivatives as multifunctional agents for the treatment of Alzheimer's disease. European Journal of Medicinal Chemistry. 2019; **178**:726-739

[65] Gan X, Hu D, Chen Z, Wang Y, Song B. Synthesis and antiviral evaluation of novel 1,3,4-oxadiazole/thiadiazole-chalcone conjugates. Bioorganic & Medicinal Chemistry Letters. 2017;**27**:4298-4301

[66] Raj R, Saini A, Gut J, Rosenthal PJ, Kumara V. Synthesis and in vitro antiplasmodial evaluation of 7-chloroquinoline-chalcone and 7-chloroquinoline-ferrocenyl chalcone conjugates. European Journal of Medicinal Chemistry. 2015;**95**:230-239

[67] Katila P, Shrestha A, Shrestha A, Shrestha R, Park PH, Lee ES. Introduction of amino moiety enhances the inhibitory potency of 1-tetralone chalcone derivatives against LPS-stimulated reactive oxygen species. Bioorganic Chemistry. 2019;**87**: 495-505498

[68] Lakshmi CSN, Balachandran S, Arul DD, Ronaldo AA, Hubert JI. DFT analysis on spectral and NLO properties of (2E)-3-[4-(dimethylamino) phenyl]-1-(naphthalen-2-yl) prop-2-en-1-one; a d-π-A chalcone derivative and its docking studies as a potent hepatoprotective agent. Chemical Data Collections. 2019;**20**:100205

[69] Kumar S, Saini A, Gut J, Rosenthal PJ, Raj R, Kumar V. 4-Aminoquinoline-chalcone/-*N*-acetylpyrazoline conjugates: Synthesis and antiplasmodial evaluation. European Journal of Medicinal Chemistry. 2017; **138**:993-1001

[70] Park CS, Ahn Y, Lee D, Moon SW, Kim KH, Yamabe N, et al. Synthesis of apoptotic chalcone analogues in HepG2 human hepatocellular carcinoma cells. Bioorganic & Medicinal Chemistry Letters. 2015;**25**:5705-5707

[71] Parambi DGT, Oh JM, Baek SC, Lee JP, Tondo AR, Nicolotti O, et al. Design, synthesis and biological evaluation of oxygenated chalcones as potent and selective MAO-B inhibitors. Bioorganic Chemistry. 2019;**93**:103335

[72] Hameed A, Abdullah MI, Ahmed E, Sharif A, Irfan A, Masood S. Anti-HIV cytotoxicity enzyme inhibition and molecular docking studies of quinoline based chalcones as potential non-nucleoside reverse transcriptase inhibitors (NNRT). Bioorganic Chemistry. 2016;**65**:175-182

[73] Melo WCMA, Santos MBD, Marques BC, Regasini LO, Giannini MJSM, Almeida AMF. Selective photoinactivation of *Histoplasma capsulatum* by water-soluble derivatives chalcones. Photodiagnosis and Photodynamic Therapy. 2017;**18**:232-235

[74] Sashidhara KV, Avula SR, Mishra V, Palnati GR, Singh LR, Singh N, et al. Identification of quinoline-chalcone hybrids as potential antiulcer agents. European Journal of Medicinal Chemistry. 2015:89638-89653

[75] Suresh S, Bhuvanesh N, Prabhu J, Thamilselvan A, Rex Jeya Rajkumar S, Kannan K, et al. Pyrene based chalcone as a reversible fluorescent chemosensor for Al^{3+} ion and its biological applications. Asian Pacific Journal of Tropical Biomedicine. 2017;**7**(8):675-679

[76] Syahri J, Yuanita E, Nurohmah BA, Armunanto R, Purwono B. Chalcone analogue as potent anti-malarial compounds against *Plasmodium falciparum*: Synthesis, biological evaluation, and docking simulation study. Asian Pacific Journal of Tropical Biomedicine. 2017;7(8):675-679

[77] Beteck RM, Legoabe LJ, Isaacs M, Hoppe HC. In vitro anti-trypanosomal activities of indanone-based chalcones. Drug Research. 2019;**69**(06):337-341

Chapter 10

Role of cAMP Homeostasis in Intra-Macrophage Survival and Infectivity of Unicellular Parasites like *Leishmania*

Arunima Biswas, Anindita Bhattacharjee and Pijush K. Das

Abstract

Unicellular eukaryotic pathogen *Leishmania donovani*, an intra-macrophage protozoan parasite, on exposure to phagolysosome conditions (PC) of mammalian macrophages, show increased cAMP level and cAMP-dependent protein kinase A (PKA) resulting in resistance to macrophage oxidative burst. In order to have a comprehensive understanding of cAMP signaling and their contribution to infectivity, studies were carried out on all the enzymes associated with cAMP metabolism such as adenylate cyclase, phosphodiesterase, pyrophosphatase and the regulatory and catalytic subunits of PKA. This chapter deals in detail the contribution of these components of cAMP signaling in cAMP homeostasis of the parasite as well as their role on successful host-parasite interaction leading to intracellular parasite survival and establishment of infection. Finally, a discussion is made about how these observations might be exploited for developing drug candidates targeting parasite specific features.

Keywords: *Leishmania*, parasite, cAMP, phosphodiesterase, pyrophosphatase, receptor adenylate cyclase, infectivity

1. Introduction

Leishmaniasis, caused by protozoan parasite *Leishmania* is still endemic in many countries and is considered as one of the potent neglected tropical disease. There are three main forms of leishmaniases-visceral (also known as kala-azar and the most serious form of the disease), cutaneous (the most common), and mucocutaneous form of the disease. Though there are surveillance and control measures for leishmaniasis being used by the World Health Organization, the treatment regime of the disease is yet to be enough to eradicate the disease worldwide. There are continuous research on potential new treatments and possible vaccines for leishmaniasis, but adequate treatment is still unavailable.

Unicellular eukaryotic pathogen *Leishmania donovani*, when exposed to phagolysosome conditions (PC) of macrophages (37°C and pH 5.5); a pre-requisite for parasite survival and infectivity, showed to elevate cAMP level and cAMP-mediated protein kinase A (PKA) activation. In eukaryotes, several researches indicate that most of the cAMP mediated effects are due to the activation of the cAMP-regulated protein kinase A, and the subsequent phosphorylation of other substrates of PKA

IntechOpen

which act as transcription factors, or metabolic enzymes such as lipases, phosphorylase kinase or glycogen synthase. In unicellular eukaryotes, there are many reports which implicates cAMP as one of the major environmental sensing machineries associated with stress response in *Plasmodium*, *Trypanosoma*, *Toxoplasma* and others. In malarial parasite, *Plasmodium falciparum*, cAMP is one of the main molecules responsible for the formation of sexual precursor, gametocytes from the asexual forms [1]. *P. falciparum* produces its own cAMP requirement by receptor adenylate cyclase (AC) which seemed to be unaffected by the well-known mammalian RAC activator Forskolin or heteromeric G-protein activators fluoroaluminate ($A1F_4^-$). Moreover, cAMP signaling effector molecule protein kinase A (PKA) plays an important role in conductance of anions across the host cell membrane of *Plasmodium*-infected RBC [2]. Moreover, recent researches showed that PKAR (PKA regulatory subunit) is functionally associated with the activation of anion conductance channel in *P. falciparum*-infected RBC [3]. cAMP-dependent signaling pathway activation and PKC activation in *Entamoeba histolytica* triggers the phosphorylation of proteins involved in actin rear-arrangements necessary for its movement and adhesion. Moreover, cAMP-response elements could play an important role in regulating actin expression and organization in signaling processes activated during tissue invasion. However, there are several other reports of mechanisms of cAMP action, such as the direct regulation of ion channels in olfactory cells, or the activation of chemotactic receptors in the slime mould, *Dictyostelium*. In unicellular eukaryotes like *Toxoplasma gondii*, both cyclic GMP (cGMP) and cyclic AMP (cAMP) can induce bradyzoite formation. These effects could be due to an increase in host or parasite cyclic nucleotides. Host cell environments including cAMP elevations contribute to the bradyzoite differentiation process in *T. gondii*, which has a receptor or sensor for cyclic nucleotides [4]. In *Dictyostelium*, cAMP secreted into the environment binds to cAMP receptors to regulate the differentiation program of cells within the fruiting body [5]. In *Leishmania*, the mechanism of action of cAMP signaling represent a particularly intriguing question since the major pathway of cAMP signaling in eukaryotes, the regulation of transcription, does not seem to be applicable because kinetoplastid parasites like *Trypanosoma* and *Leishmania* exhibit obscure transcriptional regulation. An attempt to understand cAMP signaling in *Leishmania* was undertaken by Seebeck and his group and initial studies in *L. major* where they identified five PDE genes, PDEA, PDEB1, PDEB2, PDEC and PDED encoding class I enzymes similar to those found in higher eukaryotes [6].

The protozoan parasite *Leishmania donovani*, when exposed to stress condition in the mammalian macrophages, encounter an oxidative burst as the first line of defense, offered by the macrophages by producing reactive oxygen species and reactive nitrogen intermediates [7, 8]. Still, a subset of the parasites can survive and transforms into amastigotes leading to disease manifestation [9, 10]. In *Leishmania*, cAMP is one of the major players driving the transformation of the parasite from promastigotes to amastigotes and allowing survival of parasites in macrophages [11]. Not only in the differentiation of *Leishmania*, cAMP also an important role in the differentiation of *Trypanosoma* from slender form to short stumpy form [11]. In kinetoplastid parasite *Trypanosoma*, cAMP levels are modulated all through the different stages of the cell cycle plays a significant function in transformation from slender forms to stumpy forms [12]. Also a stumpy induction factor (SIF) has been reported in *Trypanosoma* which triggers cell cycle arrest in G_1/G_0 phase and induces differentiation with high efficiency and elicits an immediate two- to three-fold elevation of intracellular cAMP content upon addition to slender forms [13]. Membrane-permeable derivatives of cAMP or the phosphodiesterase inhibitor etazolate perfectly mimic SIF activity in *Trypanosoma*. Moreover, it was also shown that the transformation in *Trypanosoma* was not mediated directly by cAMP or cAMP-analogs but by the products of hydrolysis of the membrane permeable cAMP-analogs [14]. In *Leishmania*, previous studies also showed that cAMP causes G_1 arrest in cell cycle which perhaps aids the parasite transformation [15]. Although cAMP seemed to induce cell cycle arrest in *Leishmania*, little is known about the intricate mechanism of the arrest. Though spatiotemporal regulation of cAMP and slight changes of it seemed important in the parasite, scanty data exist regarding the potential toxicity of *Leishmania* cells to pharmacologic elevation of cAMP levels. Moreover, in several mammalian systems, elevation of cAMP level is one of the stimuli that can induce growth arrest or cell death (or both) in many cultured lymphoid cells, including resting B cells, germinal center B cells, T lymphocytes, and thymocytes [16–20]. cAMP also induces cell death in cells derived from lymphoid malignancies, including murine lymphoma cell line S49.1, B-CLL cells, and multiple myeloma cells [21, 22].

To understand the importance of canonical cAMP signaling components, enzymes associated with cAMP metabolism were studied. cAMP is universally generated by adenylate cyclase in a G-protein coupled receptor signaling cascade, which catalyzes the cyclization of ATP to cAMP. In *Leishmania*, the absence of G-proteins made this signaling cascade a unique one. In many instances, adenylate cyclase is regulated by various molecules including bicarbonate, calcium, and hormones. Interestingly, our studies confirmed the importance of inorganic pyrophosphate pool (PPi), an energy storage compound and byproduct of cAMP synthesis, as one of the regulators of receptor adenylate cyclases in *Leishmania*. Also, amongst the stage specific receptor adenylate cyclases, LdRAC-A showed to regulate cAMP levels in the parasite when exposed to phagolysosome conditions. The PPi pool seemed to a stringent control by membrane bound pyrophosphatases of acidocalcisomes (ACms). Downstream, a differentially expressed soluble cytosolic cAMP phosphodiesterase (LdPDEA) and another cytosolic cAMP-dependent PDE, LdPDED, seemed responsible for controlling cAMP homeostasis. Also, a functional cAMP-binding effector molecule from *L. donovani* (a regulatory subunit of PKA, LdPKAR) seemed important in parasite infectivity playing a substantial role in autophagy induction, an event important for parasite transformation in phagolysosome conditions. Protein phosphorylation in a cAMP-dependent manner is important in the life cycle of the parasite and in any trypanosomatids, the pattern of protein phosphorylation changes within the life cycle of the parasite [23–32].

This chapter will deal in detail, the components of cAMP signaling in the parasite and unequivocally demonstrate their contribution in cAMP homeostasis; an important event for parasite survival, successful host-parasite interaction, which might be exploited for developing drug candidates targeting parasite specific features.

2. cAMP and associated enzymes in *Leishmania*

In eukaryotes, cAMP a second messenger, is an essential molecule playing a vital role in intracellular signaling which control a vast array of cellular events like cytoskeletal modeling, proliferation, virulence, differentiation and apoptosis [33]. cAMP is formed from adenosine triphosphate (ATP) by receptor adenylate cyclases (RAC). In *Leishmania*, there are reports of several isoforms of both membrane bound receptor adenylate cyclases [34] as well as soluble adenylate cyclases. When cAMP is produced, inorganic phosphate (Pi) is also produced as one of by-product of the reaction. Regulation of the pyrophosphate (PPi) pool formed by the accumulation of Pi, is hydrolysed by pyrophosphatase. In *Leishmania*, there are three isoforms of pyrophosphatases: Inorganic pyrophosphatase (IoPPase), vacuolar proton transporting pyrophosphatase (V-H^+PPase) and acidocalcisomal soluble pyrophosphatase (VSP1). Downstream to cAMP, leishmanial phosphodiesterases (PDE) hydrolyzes cAMP to 5′ adenosine monophosphate (5′AMP). There are five different PDEs in the parasite (PDEA, PDEB1, PDEB2, PDEC, and PDED). cAMP-dependent protein kinase A (PKA) exists as an inactive tetramer consisting of two catalytic subunits (PKAC) and two regulatory subunits (PKAR). Binding of cAMP to PKAR releases PKAC subunit.

2.1 Receptor adenylate cyclase in *Leishmania*

cAMP signaling cascade is activated only when local cAMP concentration reaches a level high enough to activate a cAMP responsive respective effector protein/s. It has been observed that mostly, the activation threshold lies around 1 ± 10 mm. The increase of cAMP from a basal level can be brought about either by the activation of one or several RACs, or by the inactivation of the PDEs. In eukaryotic cells, cAMP is predominantly generated at the plasma membrane since most of the known RACs are integral membrane proteins. From the site of its generation, the cAMP diffuses until it hits the respective effector molecule, or until it is hydrolysed by PDEs (**Figure 1**). The cAMP signal can take the form of a diffusion-controlled concentration gradient [35], it can be delivered in the form of time- and space-controlled spikes of cAMP concentration or consists of a sustained increase or decrease in intracellular cAMP concentration. Adenylate cyclase-cAMP pathway is also involved in the internalization process of the parasite by the host cells [36].

Studies have confirmed that cAMP is involved in signal transduction events occurring during transformations in *Leishmania* and other related kinetoplastid protozoa. Different life cycle stages contain different intracellular concentrations of cAMP in *Trypanosome brucei* [37] and in *T. cruzi* [23]. Furthermore, cAMP analogs and phosphodiesterase inhibitors promote *in vitro* differentiation of non-infectious epimastigotes of *T. cruzi* into infectious metacyclic trypomastigotes [38]. This major

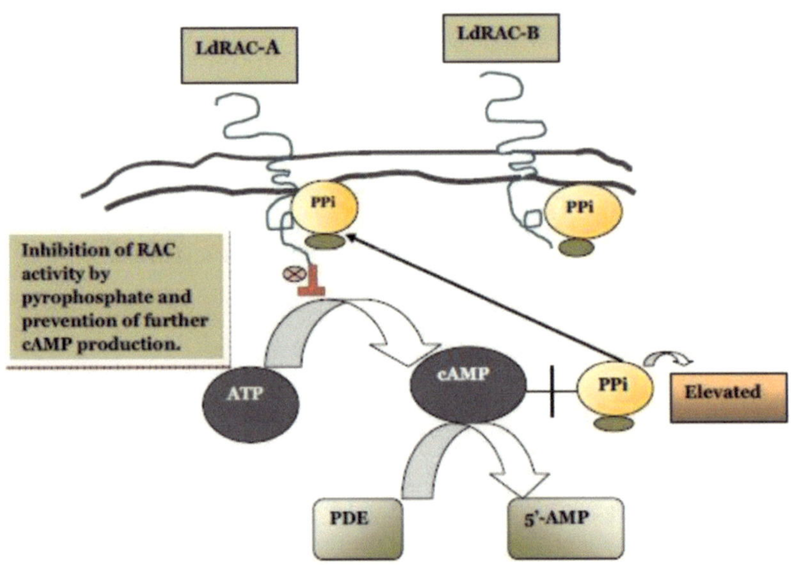

Figure 1.
Receptor adenylate cyclase in Leishmania*: PPi inhibiting adenylate cyclase. In normal condition, when the parasites are not exposed to stress, receptor adenylate cyclase synthesizes cAMP from ATP and PPi is produced as by-product. This PPi interacts with the receptors and inhibits further synthesis of cAMP. On the other hand, PDEs present in the cell also helps in maintaining the concentration of cAMP by hydrolyzing cAMP to 5′-AMP. So, both PPi produced and the PDEs help in the regulation of cAMP level in the parasite.*

role of cAMP in the transformation of kinetoplastid protozoa has led to the investigation of adenylate cyclases in *L. donovani*. A family of five clustered genes in *L. donovani* was identified which encodes signal transduction receptors [39]. The coding region of these genes was sequenced and have been shown to code for proteins with an NH_2-terminal hydrophilic domain, an intervening transmembrane segment and a carboxylic terminal domain having high sequence similarity with the catalytic domain of adenylate cyclases from other eukaryotes [39]. These genes are designated as rac-A and rac-B. One of these genes is expressed in *Xenopus* oocytes and have been shown to function as an adenylate cyclase. Another interesting observation was that of rac-A and rac-B mRNAs expression which was only found in promastigotes by Northern blot technique but was not detectable in amastigotes, which proves that these are developmentally regulated mRNAs [39]. So, these proteins might be involved in developmental transitions where they can function in the switch between non-infectious procyclic and infectious metacyclic promastigotes [40], or they can also interact with ligands present in the host macrophages and initiate a signal cascade leading to the differentiation of promastigotes to amastigotes.

In *T. brucei*, genes showing homology with yeast adenylate cyclase were identified and they were termed expression-site associated genes (ESAGs). Many more copies of these putative adenylate cyclases were identified and were named GRESAG4.1 and GRESAG4.2 [41]. Related adenylate cyclase genes which have actually been proved to code for functional adenylate cyclase enzymes were also identified in *T. congoense*, *T. mega*, *T. brucei gambiense*, *T. vivax* and *T. equiperdum* [42, 43]. Similar families of multigene having high homology with ESAG4 and GRESAG4.1 have been identified in *T. cruzi* and *L. donovani* are said to be sharing the common protein architecture [39]. In kinetoplastids, the cellular localization of adenylate cyclases is consistent where they act as receptors as proved by binding of antibodies against ESAG4, specifically to the cell surface along the flagella in trypanosomes [43].

The existence of receptor adenylate cyclase has also been discovered in *L. donovani* and a membrane bound RAC-A is found to be functional during exposure to phagolysosome condition (PC) which actively catalyze cAMP generation [39]. Expression of receptor adenylate cyclase mRNAs (RAC-A and RAC-B) was also found to be developmentally regulated in *Leishmania* as their expression was only found in promastigotes but not in amastigotes [39]. It has been reported that promastigotes exposed to PC shown elevated level of cAMP after 60 minutes of PC exposure which was decreased when treated with DDA (di-deoxy adenosine), an adenylate cyclase inhibitor [44]. Expression of both LdRAC-A and LdRAC-B were analyzed by immunoblot technique using anti-RAC-A and anti-RAC-B antibodies raised against *Leishmania* and their expressions were revealed in both plasma membrane and flagella. Interestingly, the expression of LdRAC-A increased significantly in PC-exposed cells after 60 minutes of exposure despite the unchanged expression profile of LdRAC-B under such condition. This result suggests that in spite of the presence of two developmentally regulated isoforms of adenylate cyclases in *Leishmania*, LdRAC-A is only functionally active during stress condition. Inducible anti-sense knock-down strategy was adopted to downregulate RAC-A and RAC-B in *L. tarentolae*, which had been successfully implemented for a number of genes earlier. Ldrac-A knocked-down parasites generated in Lt.T7TR strain of *L. tarentolae* showed no change or no elevation in intracellular cAMP level after exposure to PC. Moreover, there was a 10.2% decrease in cAMP level when RAC-A knocked-down PC-exposed parasites were further treated with DDA. Ldrac-B knocked-down parasites behaved as control parasites showing elevated cAMP levels on PC exposure. However, there was a significant decrease of cAMP level when RAC-B knocked-down cells were treated with DDA. The results indicate that LdRAC-A plays a conspicuous role in triggering cAMP response in the parasites during stress condition.

2.2 Enzymes regulating receptor adenylate cyclase function in *Leishmania*: pyrophosphatase

Pyrophosphates (PPi) are produced as by-product during the conversion of ATP to cAMP by receptor adenylate cyclase, the product accumulation of which inhibits adenylate cyclase reaction toward the formation of cAMP. PPi is found to be stored in a specialized compartment like acidocalcisomes in kinetoplastid parasites [45]. The concentration of PPi is equivalent to that of ATP in the cell in spite of its huge confinement in the acidocalcisomes of *Leishmania*. There were speculations that this high concentration of PPi (in millimolar range) might be responsible for the inhibition of cAMP production in the parasites by modulation adenylate cyclase reaction in subcellular micro domains [46]. There are at least three different pyrophosphatases present in *L. major* as revealed by genome sequence analysis. These are: membrane associated H^+-translocating pyrophosphatase (V-H^+PPase), soluble acidocalcisomal pyrophosphatase (VSP1) and an inorganic pyrophosphatase (IoPPase) and are responsible for maintaining the cAMP levels in the parasite.

Further studies have been conducted to elaborately decipher the role that RAC plays along with various molecules associated with it. PPi formed as by-product of cAMP biosynthesis inhibits adenylate cyclase function and this inhibition is reversed when PPi is hydrolysed by acidocalcisomal LdV-H^+PPase which is translocated to plasma membrane on exposure to phagolysosome condition (**Figure 2**).

Apart from the direct role of LdRAC-A in the production of cAMP during stress condition, intracellular PPi and pyrophosphatases also play a major role in regulation of cAMP concentration in the cell. *L. donovani* promastigotes were treated with foscarnet, a pyrophosphate analogue that acts as an adenylate cyclase inhibitor [47] under PC-exposed condition. PC induced cAMP generation was inhibited by foscarnet treatment after 60 minutes of PC exposure [44]. Furthermore, in PC exposed cells, total pyrophosphate pool was markedly reduced. Presence of three pyrophosphatases have been detected in *L. donovani*, namely, soluble acidocalcisomal pyrophosphatases (LdVSP1), vacuolar proton transporting pyrophosphatase (LdV-H^+PPase) and inorganic pyrophosphatase (LdIoPPase) which collectively maintain the intracellular pyrophosphate pool. Co-localization analysis with cells expressing GFP-fusion proteins of the three pyrophosphatases and acidocalcisome-targeted dye DND-lysotracker, showed little localization of LdIoPPase which was localized in cytoplasm but significant co-localization was observed for LdVSP1 and LdV-H^+PPase they were predominantly localized in the acidocalcisomes.

As revealed by immune-electron microscopic analysis, the acidocalcisomes localize in the vicinity of the cell membrane on PC exposure. PC exposure resulted in gradual decrease in intraluminal pH because of enhanced proton import by LdV-H^+PPase indicating translocation of acidocalcisome that actively imports proton, in the cell periphery following PC exposure (**Figure 2**). The translocation of acidocalcisome to membrane vicinity was further explored to find the mechanism behind such stress driven translocation. Studies clearly indicated that the movement of acidocalcisomes during stress is a microtubule and microfilament-dependent process. Pre-treatment with F-actin inhibitor, cytochalasin D, and stress exposure showed absence of acidocalcisomal translocation toward membrane. Nocodazole pre-treatment, an inhibitor of microtubule, and subsequent stress exposure also resulted in inhibition of acidocalcisomal translocation [44].

Moreover, presence of putative actin/tubulin binding proteins in *Leishmania* might provide significant clues and insight on interlinking of cytoskeletal rearrangement. One such protein has been cloned from *L. donovani* (cyclase-associated protein, LdCAP1) (Bhattacharya et al. personal communication). Unravelling the function of the same might throw light on cytoskeletal protein rearrangement

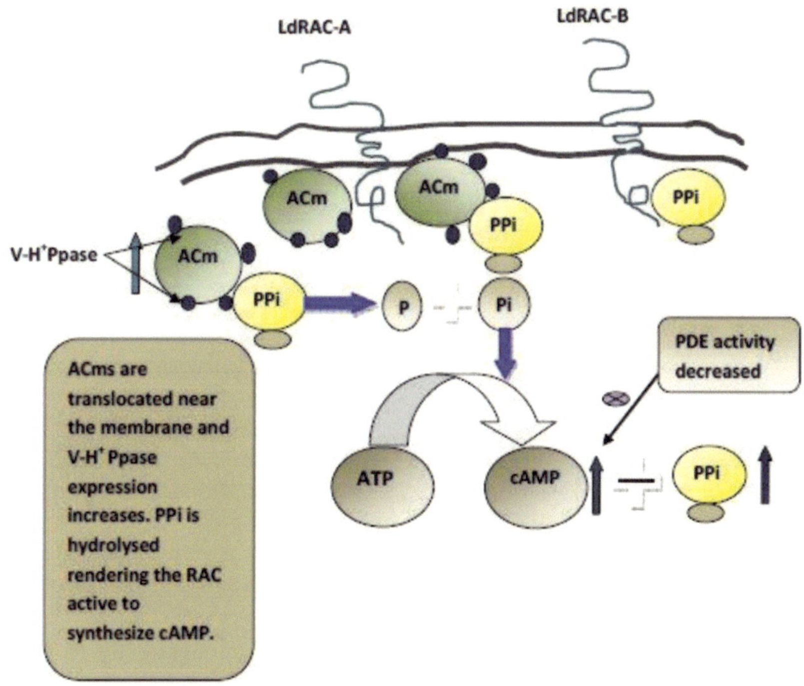

Figure 2.
Role of receptor adenylate cyclase when parasites are exposed to phagolysosomal conditions. Generally PPi inhibits adenylate cyclase function in normal conditions, but when cells are exposed to stress, the PPi are hydrolysed by acidocalcisomal V-H$^+$PPase and translocated toward membrane vicinity allowing receptor adenylate cyclase to synthesize more cAMP. In addition to this, PDE level decreases in stress-induced cells which also elevates cAMP level.

and acidocalcisomal translocation. With the translocation of acidocalcisomes in membrane vicinity, a possible co-proximal localization of membrane bound acidocalcisome (LdV-H$^+$PPase) and LdRAC-A was studied during PC exposure. No such co-localization was detected with LdRAC-B. Episomal over-expression and conditional silencing demonstrated regulatory role of V-H$^+$PPase on cAMP production. Though the direct decrease in the level of PPi by V-H$^+$PPase could not be established by the study of Biswas et al. [44], the use of PPi analogue foscarnet and the decrease in the PPi level during PC exposure indicate toward the regulation of PPi pool by this pyrophosphatase isoform. LdRAC-A, PPi pool and LdV-H$^+$PPase control intracellular cAMP level in the parasite during PC exposure.

2.3 Phosphodiesterases and intracellular cAMP signaling in *Leishmania*

Apart from pyrophosphatases that regulate the formation of intracellular cAMP by receptor adenylate cyclases, it is also important to study another dimension of cAMP regulation. Phosphodiesterases (PDEs), ubiquitous enzymes responsible for the termination of cyclic nucleotide signaling pathway by hydrolyzing cAMP to 5′-AMP or cGMP to 5′-GMP, the sole means by which the cell gets rid from the cAMP produced for controlling different cellular processes [48]. PDEs can be divided into three categories based on their catalytic properties namely, class I, class II and class III and 21 genes have been found in mammals for PDE and several in *Drosophila* and *Dictyostellium*. Though various isoforms of class I PDE have been identified in *T. brucei* and *T. cruzi*, only two PDEs have been cloned from *L. major* [48]. In *L. major* five different isoforms of PDE have been identified. Isoforms PDEB1 and PDEB2 are highly specific for cAMP and only poorly inhibited by

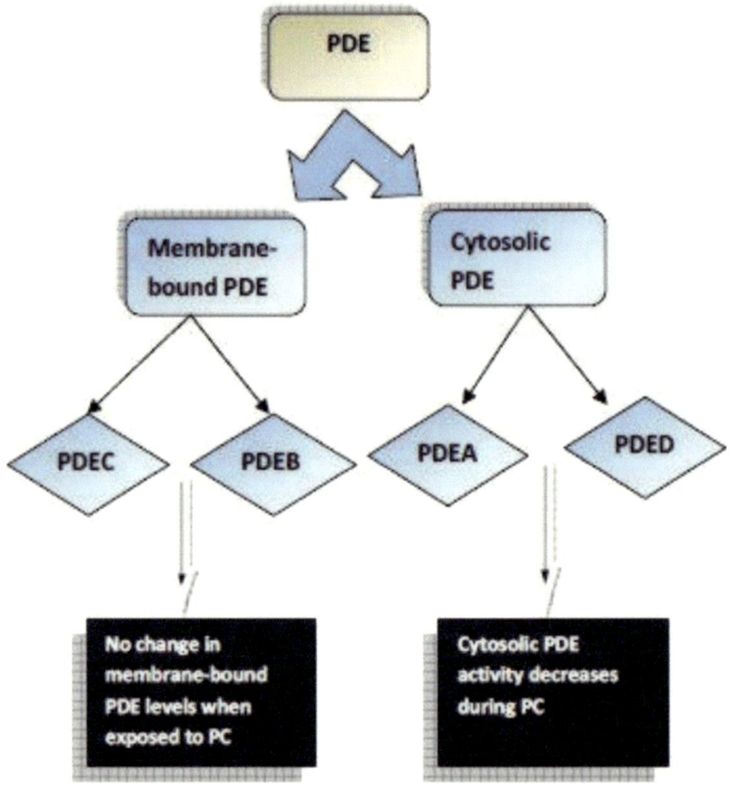

Figure 3.
PDE isoforms in Leishmania*. PDE in* Leishmania *can be categorized into membrane-bound and cytosolic PDE. On exposure to phagolysosome condition, there is no change in the expression of membrane-bound PDEs (PDEC and PDEB); but there is a significant decrease in the expression of cytosolic PDEs (PDEA and PDED).*

most inhibitors of human PDEs [48]. Crystal structure of LmjPDEB1 showed that catalytic domain of LmjPDEB1 complexed with a general PDE inhibitor, 3-isobutyl-1-methyl-xanthine (IBMX) show significant differences in binding of this inhibitor when compared to human PDEs.

Identification of different isoforms of phosphodiesterases in *L. major* indicated PDEB and PDEC to be membrane bound and PDEA and PDED to be predominantly cytosolic (**Figure 3**). LdPDEA and LdPDED were also cloned in *L. donovani* [49]. From the observation of the studies of Bhattacharya et al. [49], it has been found that the activity of cytosolic PDEs decrease during stage differentiation but the activity of membrane bound PDEs remained unchanged. From this observation, it can be inferred that PDEs might play an essential role as a controlling factor during stage differentiation of the parasites.

When cAMP-PDE activity was studied, it was found that the activity of cytosolic fraction was diminished gradually as the parasite started to differentiate into axenic amastigote stage from log phase promastigote. Protein level expression of different forms of PDEs in different stages of life cycle of *L. donovani* revealed depletion of PDEA expression in late stationary-phase promastigotes and axenic amastigotes as compared to log phase promastigotes but the expression of other PDEs such as PDEC and PDEB remain unaltered. Gradual decrease in PDEA level and its differential expression in the course of the differentiation of the parasites from promastigotes and amastigotes was observed by several experimental techniques.

2.3.1 Effect of PDEA on peroxide resistance and TSH pool

In *Leishmania*, anti-oxidant machinery plays a vital role in regulating the sustenance of the parasites in mammalian macrophages where they are exposed to oxidative stress. cAMP level elevation is linked with such phenomenon. In order to find out the functional significance of LdPDEA in such defense mechanism, LdPDEA gene was silenced using tetracycline-inducible knock-down system [49]. When PDE inhibitors were used, the parasites exhibited enhanced viability against peroxides and peroxynitrite. When cells were treated with PDE inhibitors like etazolate and trequinsin, higher resistance against peroxide and peroxynitirite was observed as compared to untreated promastigotes. Since these inhibitors are not specific for PDEA, the result of the treatment might be due to inhibition of some other forms of PDEs in the promastigotes. To ascertain the exact role of PDEA, a knock down construct was prepared to build up a tetracycline-inducible PDEA knock down system. PDEA expression was strongly reduced in both RNA and protein level after tetracycline induction and they also showed enhanced resistance against peroxide and peroxynitrite.

Peroxide neutralization is one of the major strategies of leishmanial parasite, which makes their survival possible inside the mammalian macrophage and it is done by anti-oxidant machinery of the parasite which lacks catalase. In *Leishmania*, peroxide neutralization is mainly based on trypanothione (TSH), a glutathione-spermidine conjugate, as they lack glutathione (GSH). TSH is biosynthesized from

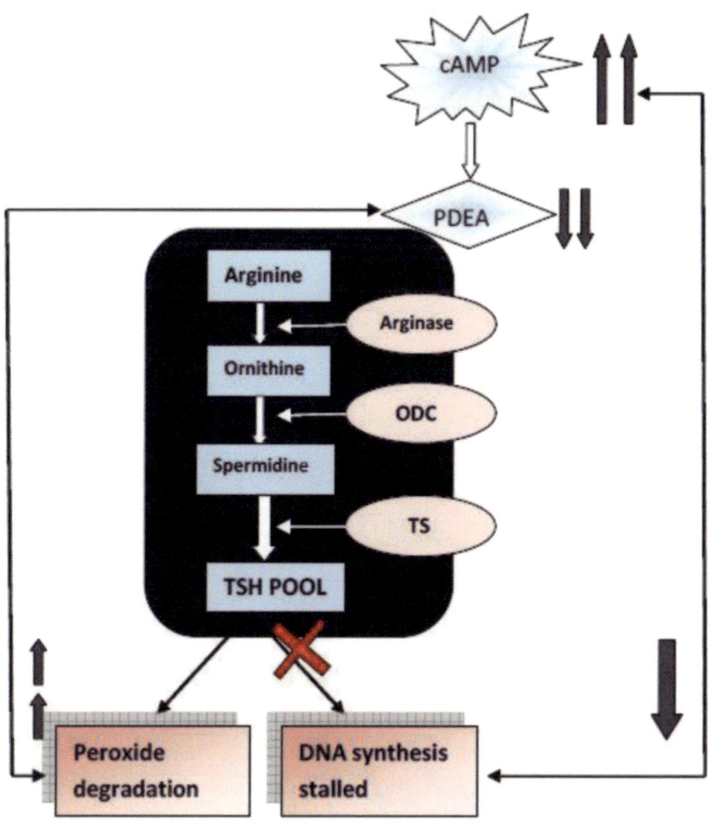

Figure 4.
Role of PDEA to control local cAMP gradient and shifting of TSH pool to peroxide neutralization.

arginine by arginase, ornithine decarboxylase and other enzymes, which converts it into spermidine and is then conjugated with GSH. No significant change in arginine and ornithine transporter was detected in PDE inhibitor treated cells and also in PDEA knocked down cells. On the contrary, when the expression of arginase and ornithine decarboxylase, the enzymes responsible for TSH biosynthesis was checked in control and PDEA inhibitor-treated cells, an increase in the expression of these enzymes was observed indicating that PDEA inhibition might have a role in TSH biosynthesis. When total thiol or intracellular TSH content was analyzed, not much alteration was observed. TSH pool is generally utilized by the parasite either for DNA replication by ribonucleotide reductase or for peroxide degradation by peroxidoxin, ascorbate peroxidase and superoxide dismutase. The expressions of enzymes responsible for peroxide degradation like peroxidoxin, superoxide dismutase and ascorbate peroxidase were elevated in PDEA-inhibited cells (**Figure 4**). Cells overexpressing PDEA also showed reduced resistance to pro-oxidants when exposed to phagolysosome condition as compared to normal cells [49].

2.3.2 Role of PDED in cAMP homeostasis

Apart from the membrane bound phosphodiesterases, a soluble, cytosolic phosphodiesterase (PDED) was cloned and characterized from *L. donovani*. Bioinformatic studies showed the presence of two pseudo-substrate sites and a putative PKA phosphorylation site at the C-terminus of PDED and PKA-mediated phosphorylation is important for the regulation of phosphodiesterase activity (**Figure 5**) [50]. It was observed that catalytic subunits of PKA (PKAC1 and PKAC2) interacts with the pseudo substrate sites of PDED after 3 hours of PC exposure. Moreover, inhibition of phosphodiesterase activity through PKA-mediated phosphorylation was observed at a further later time point of PC exposure [51]. The cytosolic localization of LdPDED was established by immunolocalization analysis using anti-LdPDED antibody which revealed its localization to be predominantly cytosolic. Interaction of LdPDED with the catalytic subunits of LdPKA within

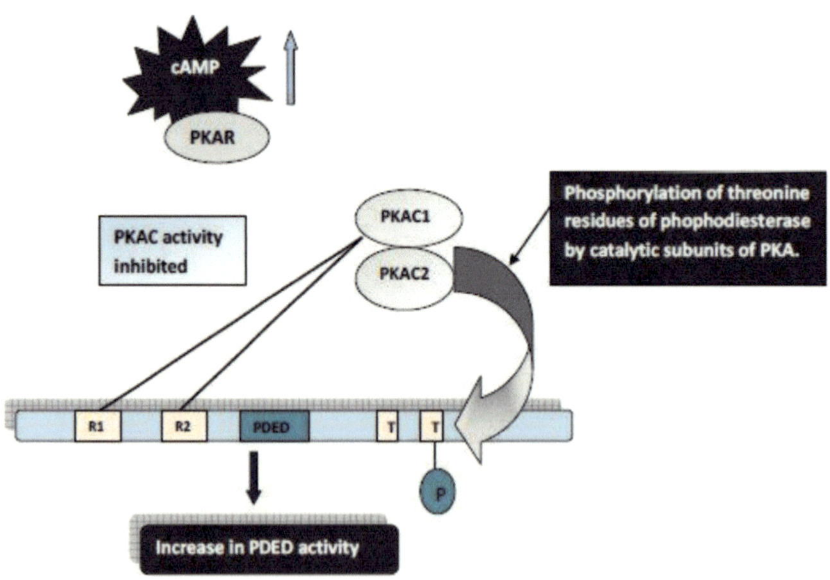

Figure 5.
LdPDED interacting with PKAC1 and PKAC2 resulting in the inhibition of kinase activity of PKA. PKA on the other hand phosphorylates threonine residue of PDED increasing its phosphodiesterase activity.

3 hours of exposure to differentiation condition leads to the inhibition of LdPKA (short-term regulation). LdPKA-mediated phosphorylation of LdPDED is observed when parasites are exposed to differentiation condition for more than 6 hours. Hydrolytic property of LdPDED is enhanced due to this phosphorylation event and this enhancement in hydrolytic activity might play a pivotal role in the maintenance of cAMP homeostasis (long term regulation) when the total cytosolic PDE activity falls because of PDEA depletion during stress condition [49]. This role of PDED in maintaining the PKA activity which in turn regulates cAMP homeostasis in the parasite during initial exposure to stress condition, might be important in the life cycle of the parasite particularly in the infection establishment within the host.

2.4 PKA as the downstream effector of cAMP in *Leishmania*

Though the existence and functioning of cAMP-dependent protein kinase (PKA) is well pronounced in eukaryotes, very little is known about the functioning of PKA in cAMP signaling of this particular parasite. PKA acts as the immediate downstream effector of cAMP in the adenylate cyclase pathway, catalyzing the transfer of γ-P from ATP to specific serine/threonine residues on the substrate protein [52]. Studies on *S. cerevisiae* reveal that one of the three PKA catalytic subunits mediates stress-induced differentiation [53]. Researches in *Dictyostelium* have suggested that cAMP is not required for differentiation if sufficient levels of PKA activity are present [54, 55] indicating profound role of PKA in differentiation. Activation of PKA by a short-term cAMP pulse induces bradyzoite differentiation, whereas a prolonged cAMP pulse inhibits differentiation [56]. It is likely that there are distinct PKA signaling pathways in the tachyzoite with opposing effects on parasite differentiation. Inhibition of PKA signals by treatment with PKA catalytic subunit inhibitor H89 induces bradyzoite differentiation [57], suggesting that PKA catalytic subunit activity may be involved in cAMP-mediated tachyzoite maintenance.

When *Leishmania* parasites were exposed to stress condition, PKA activity was significantly enhanced along with increased level of cAMP. Protein kinase activity of five different species of *Leishmania* was found to be quite high in both logarithmic and stationary phase promastigotes, being most active in *L. amazonensis* and least in *L. donovani* [58]. PKA catalytic subunits in the *Toxoplasma* genome were identified. PKA is the most important downstream effectors of cAMP signaling pathway and it exists as an holoenzyme in inactive state with the association of regulatory subunit [59–61]. In case of cAMP analog-treated cells and PC-exposed cells, substrate level phosphorylation on serine and threonine residues were also found to be increased. In most of the eukaryotic cells, PKA exist as an inactive tetrameric holoenzyme consisting of two catalytic and two regulatory subunits denoted as PKA-C and PKA-R respectively. The PKA-R subunit actually binds with cAMP causing a conformational change in the molecule resulting in the dissociation of the R and C subunits of the holoenzyme. This dissociation activates the catalytic C subunit of PKA which phosphorylates specific serine or threonine residues on substrate proteins in the cytoplasm and nucleus [62].

A 34 KD protein with similar properties of mammalian PKA-C was purified from *L. donovani* [63]. The effect of different activators and inhibitors on PKA activity was measured using promastigote lysates and fluorescent kemptide and it was found that though cAMP analogue treatment did not have any conspicuous effect on kemptide phosphorylation, treatment with PKA inhibitors like PKI and H89 profoundly decreased kemptide phosphorylation. On the other hand, PDE-resistant PKA activators increased kemptide phosphorylation when compared to basal activity. Addition of PDE inhibitors like dipyridamole and rolipram also increased kemptide phosphorylation [64]. These results suggest that cAMP has some direct role in the activation of PKA during transformation in *Leishmania*. Treatment of promastigotes with PKA activators also resulted in growth arrest in the parasite [64]. Parasite survival in the peritoneal macrophages of Balb/c mice was examined using PKA-inhibitor treated parasites and there was a significant reduction in macrophage infection [64].

In spite of the discovery of the role played by adenylate cyclases and phosphodiesterases in cAMP homeostasis of *Leishmania*, existence of no specific cAMP-binding effector molecule was known. Bhattacharya et al. [65], in their studies, have identified a regulatory subunit of cAMP-dependent protein kinase (Ldpkar1) in *L. donovani* which was found to be homologous to class I cAMP-dependent protein kinase regulatory subunit of mammals. Studies proved beyond doubt that this regulatory subunit interact with both the catalytic subunits of PKA, thus inhibiting PKA activity. When co-immunoprecipitation assay was performed for both normal and Sp-8-Br-cAMP-pretreated cells, much weaker signal was detected for treated cells as compared to normal cells suggesting Sp-8-Br-cAMP-mediated activation of PKA. Moreover, when activity was analyzed in LdPKAR1-LdPKAC1 and LdPKAR1-LdPKAC2 immunoprecipitated complexes in the presence or absence

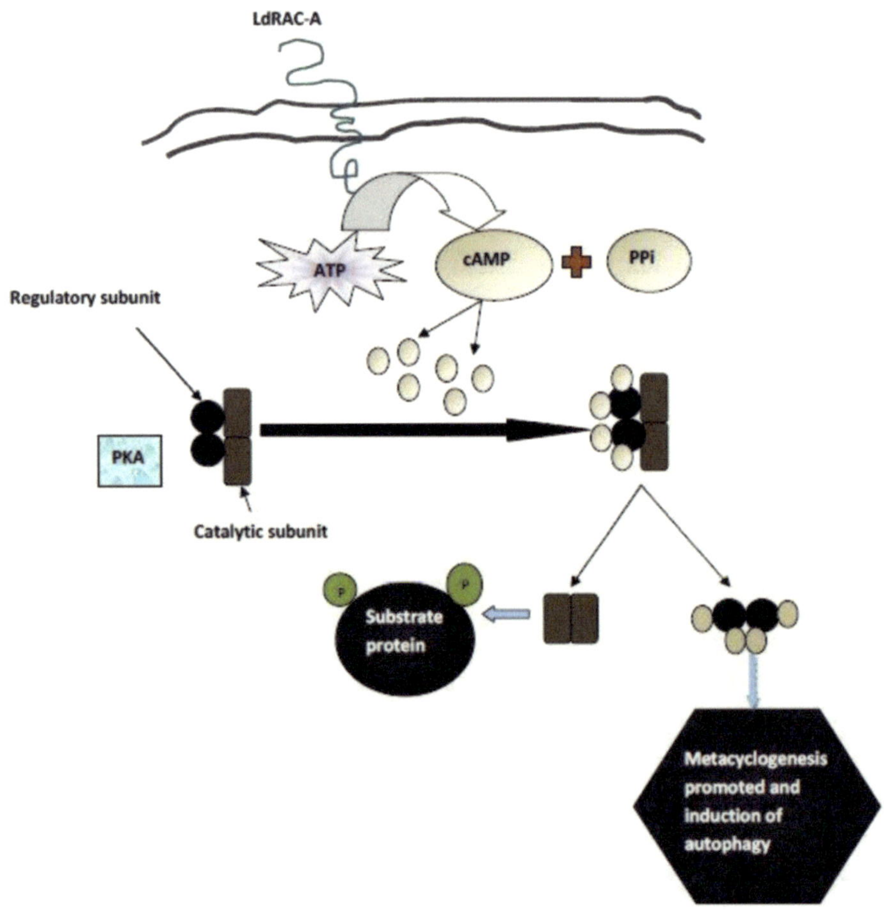

Figure 6.
cAMP-dependent PKA activity in Leishmania*: cAMP level increases on phagolysosome condition exposure and cAMP binds with the regulatory subunit of PKA enabling its dissociation from the catalytic subunit rendering the catalytic subunit active. cAMP associated regulatory subunit promotes metacyclogenesis and induces autophagy whereas activated catalytic subunit phosphorylates other proteins downstream to this signaling cascade.*

of excess cAMP, kemptide phosphorylation was increased significantly in the presence of cAMP indicating the pivotal role of cAMP in the dissociation of regulatory subunit from the catalytic subunit rendering the latter active. Moreover, the studies of Bhattacharya et al. [65] also establishes the functional importance of PKAR other than working as a cAMP effector molecule. LdPKAR1 expression was also found to be increased in late stationary phase promastigotes kept in nutrient deprived/starvation condition and metacyclogenesis, which is a pre-requisite for successful macrophage infection, was significantly induced in starved cells as compared to normal cells. Cells overexpressing LdPKAR1 also showed increased metacyclogenesis, enhanced intra-macrophage survival suggesting that LdPKAR1 overexpressed cells had greater infectivity. It can be inferred that LdPKAR1 overexpression leads to acceleration in the process of metacyclogenesis in *L. donovani* (**Figure 6**).

PKA activity assay in the presence and absence of cAMP and cGMP analogs and PKA inhibitors in both soluble fraction (SF) and membrane fraction (MF) of infective promastigotes of *L. amazonensis* showed increase in phosphorylative activity of the kinase in cAMP-analog-treated cells, and not in cGMP-analog-treated cells, was conspicuous, particularly in the SF of the promastigotes. On the contrary, PKA activity of both SF and MF of axenic amastigotes was found to be much lower as compared to that of both SF and MF of infective promastigotes under same experimental conditions [66].

Autophagy is one of the survival strategies of *Leishmania* in mammalian macrophages. Since LdPKAR1 has a direct role in the process of metacyclogenesis, its relation to autophagy was studied in the parasite. ATG8 is a marker for autophagosome formation and ATG8 tracking was done in both starved cells and in normal cells by western blot technique using polyclonal anti-LmATG8 antibody. Cells under starvation condition showed much higher level of ATG8-PE, a cleaved form of ATG8, indicating the formation of autophagosome in starved condition. When a conditional knock-down system of LdPKAR1 was constructed in *L. tarentolae*, both mRNA and protein level expression of LdPKAR1 was found to be diminished after tetracycline induction. Uninduced cells showed higher percentage of ATG8-positive structures as compared to tetracycline-induced cells. This suggested the role of PKAR1 in autophagosome formation. LdPKAR not only acts as a cAMP binding molecule in the parasite, but induce metacyclogenesis and autophagy. Studies are further required to confirm whether the process is an autophagy-induced metacyclogenesis or a metacyclogenesis-induced autophagy.

3. Conclusion

To conclude we can say that the leading researches in the recent past has enriched our knowledge on the importance of cAMP signaling in kinetoplastid parasites like *Leishmania* and their association with parasite infectivity. These findings provide insight on the functioning of different enzymes associated with cAMP metabolism (**Figure 7**). These studies point toward the fact that modulation of cAMP level in the parasite might be one of the mechanisms to control leishmaniasis and the molecules associated with the same might be tested as potent drug targets against the disease.

Presently, PDE inhibitors are potent drug targets against various human diseases. Study of human PDEs in cAMP signaling pathway has revealed their druggability in various human pathologies leading to various marketed drugs [67]. Moreover, there is a similarity between human and protozoan enzymes and in addition, the availability of human PDE inhibitors as therapeutics has thrown some light on the discovery of some specific protozoan PDE inhibitors as potential drug targets [68]. In kinetoplastid parasites like *Trypanosoma*, PDE inhibitors are being screened as potential drug targets

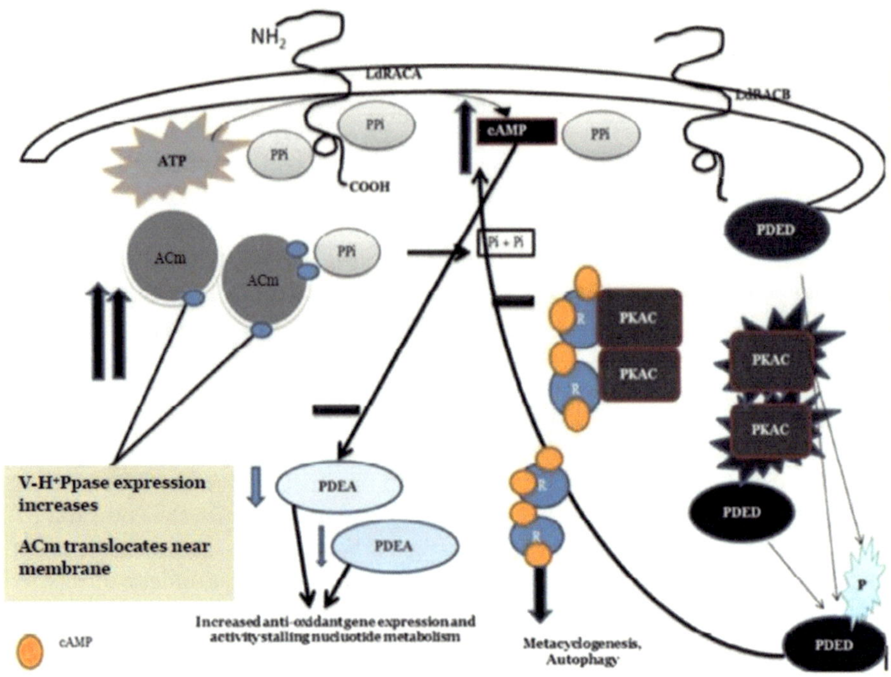

Figure 7.
An overview of cAMP signaling in Leishmania *during stress condition where receptor adenylate cyclase, PDEA, PDED, PKA the effector molecule of cAMP and acidocalcisomal pyrophosphatases play the major roles in the maintenance of cAMP homeostasis.*

and pharmacological validation of using PDEs as novel drug targets for diseases caused by the kinetoplastid parasites. This study for the first time explores the possibility of using human PDE inhibitors as the starting framework for the design of *Leishmania*-selective inhibitors. This proposal is supported by the inhibitory effects of some human PDE inhibitors observed on *T. cruzi* PDEC, e.g., etazolate inhibits human PDE4 and TcrPDEC with the IC_{50} values of 2 and 0.7 nM [69]. Among the PDE inhibitors used in this study etazolate along with rolipram showed maximum anti-proliferative activity against *Leishmania* parasite with least cytotoxic effect on macrophage cells cultured *in vitro*. Further it was observed that both of them significantly affected the G1 cell cycle arrest and mitochondrial membrane potential of the parasite, therefore we assessed *in vitro* for their ability to clear parasite load within the macrophage cells. Etazolate was found to be more effective in clearing the parasite load when the macrophage cells were pretreated with it compared to rolipram. Etazolate belongs to pyrozopolidine class compound which shows PDE4 enzyme inhibitory activity [70]. Preclinical studies as well as pharmacokinetic and safety profiles in Phase I and Phase IIa of clinical studies revealed that etazolate is a well-established drug of choice with no major side effects reported [70]. Etazolate produced antidepressant like effects in animal models of depression and at the same time it could be used in the treatment of Alzheimer's disease [71]. If through experimentation a minimal dose of etazolate could be determined then etazolate could itself be used as an anti-leishmanial drug. However, if it is not possible to determine the dose concentration which would specifically inhibit parasite PDE, then one can make use of the significant advances made in medicinal chemistry to design compounds which could specifically inhibit parasite PDE but not that of the host. This compound if developed, could act as a potential anti-leishmanial agent in future.

On the other hand, several known PDE inhibitors were tested against *Plasmodium* PDEα, and zaprinast, a known selective inhibitor of human PDE5 which is specific for both cAMP and cGMP, turned out to be the most potent, with an IC_{50} value of 3.8 μM [72]. High concentration of PDE inhibitor like dipyridamole resulted in the inhibition of promastigote proliferation and macrophage infection in *L. major* [64].

Acknowledgements

We thank DST-INSPIRE Project grant (IFA-12 LSBM-22), PRG-University of Kalyani and NASI Senior Scientist Platinum Jubilee Fellowship for this work.

Conflict of interest

The authors declare no conflict of interest.

Acronyms and abbreviations

AC	acidocalcisomes
cAMP	cyclic adenosine monophosphate
PDE	phosphodiesterase
PKA	protein kinase A
PPi	inorganic pyrophosphate
RAC	receptor adenylate cyclase

Author details

Arunima Biswas[1], Anindita Bhattacharjee[1] and Pijush K. Das[2]*

1 Department of Zoology, Cell and Molecular Biology Laboratory, University of Kalyani, Kalyani, India

2 Infectious Diseases and Immunology Division, CSIR-Indian Institute of Chemical Biology, Kolkata, India

*Address all correspondence to: pijush52@gmail.com

IntechOpen

© 2019 The Author(s). Licensee IntechOpen. This chapter is distributed under the terms of the Creative Commons Attribution License (http://creativecommons.org/licenses/by/3.0), which permits unrestricted use, distribution, and reproduction in any medium, provided the original work is properly cited.

References

[1] Kaushal DC, Carter R, Miller LH, Krishna G. Gametocytogenesis by malaria parasites in continuous culture. Nature. 1980;**286**(5772):490-492

[2] Egée S, Lapaix F, Decherf G, Staines HM, Ellory JC, Doerig C, et al. A stretch-activated anion channel is up-regulated by the malaria parasite Plasmodium falciparum. Journal of Physiology (London). 2002;**542**(Pt 3):795-801

[3] Merckx A, Nivez M-P, Bouyer G, Alano P, Langsley G, Deitsch K, et al. Plasmodium falciparum regulatory subunit of cAMP-dependent PKA and anion channel conductance. PLoS Pathogens. 2008;**4**(2):e19

[4] Kirkman LA, Weiss LM, Kim K. Cyclic nucleotide signaling in toxoplasma gondii bradyzoite differentiation. Infection and Immunity. 2001;**69**(1):148-153

[5] Loomis WF. Cell signaling during development of *Dictyostelium*. Developmental Biology. 2014;**391**(1):1-16

[6] Johner A, Kunz S, Linder M, Shakur Y, Seebeck T. Cyclic nucleotide specific phosphodiesterases of *Leishmania major*. BMC Microbiology. 2006;**6**:25

[7] Zarley JH, Britigan BE, Wilson ME. Hydrogen peroxide-mediated toxicity for *Leishmania donovani* chagasi promastigotes. Role of hydroxyl radical and protection by heat shock. The Journal of Clinical Investigation. 1991;**88**(5):1511-1512

[8] Gantt KR, Goldman TL, McCormick ML, Miller MA, Jeronimo SM, Nascimento ET, et al. Oxidative responses of human and murine macrophages during phagocytosis of Leishmania chagasi. Journal of Immunology. 2001;**167**(2):893-901

[9] Pearson RD, Sullivan JA, Roberts D, Romito R, Mandell GL. Interaction of *Leishmania donovani* promastigotes with human phagocytes. Infection and Immunity. 1983;**40**(1):411-416

[10] Pearson RD, Wheeler DA, Harrison LH, Kay HD. The immunobiology of leishmaniasis. Clinical Infectious Diseases. 1983;**5**(5):907-927

[11] Breidbach T, Ngazoa E, Steverding D. *Trypanosoma brucei*: In vitro slender-to-stumpy differentiation of culture-adapted, monomorphic bloodstream forms. Experimental Parasitology. 2002;**101**(4):223-230

[12] MacGregor P, Ivens A, Shave S, Collie I, Gray D, Auer M, et al. High-throughput chemical screening for antivirulence developmental phenotypes in *Trypanosoma brucei*. Eukaryotic Cell. 2014;**13**(3):412-426

[13] Vassella E, Reuner B, Yutzy B, Boshart M. Differentiation of African trypanosomes is controlled by a density sensing mechanism which signals cell cycle arrest via the cAMP pathway. Journal of Cell Science. 1997;**110**(Pt 21):2661-2671

[14] Laxman S, Riechers A, Sadilek M, Schwede F, Beavo JA. Hydrolysis products of cAMP analogs cause transformation of *Trypanosoma brucei* from slender to stumpy-like forms. Proceedings of the National Academy of Sciences of the United States of America. 2006;**103**(50):19194-19199

[15] Bhattacharya A, Biswas A, Das PK. Role of intracellular cAMP in differentiation-coupled induction of resistance against oxidative damage in *Leishmania donovani*. Free Radical Biology & Medicine. 2008;**44**(5):779-794

[16] Daniel V, Litwack G, Tomkins GM. Induction of cytolysis of cultured

lymphoma cells by adenosine 3′:5′-cyclic monophosphate and the isolation of resistant variants. Proceedings of the National Academy of Sciences of the United States of America. 1973;**70**(1):76-79

[17] Lømo J, Blomhoff HK, Beiske K, Stokke T, Smeland EB. TGF-beta 1 and cyclic AMP promote apoptosis in resting human B lymphocytes. Journal of Immunology. 1995;**154**(4):1634-1643

[18] Knox KA, Gordon J. Regulation of apoptosis in germinal centre B cells. Immunology Today. 1995;**16**(2):106-107

[19] McConkey DJ, Orrenius S, Jondal M. Cellular signalling in programmed cell death (apoptosis). Immunology Today. 1990;**11**(4):120-121

[20] Kizaki H, Suzuki K, Tadakuma T, Ishimura Y. Adenosine receptor-mediated accumulation of cyclic AMP-induced T-lymphocyte death through internucleosomal DNA cleavage. The Journal of Biological Chemistry. 1990;**265**(9):5280-5284

[21] Krett NL, Zell JL, Halgren RG, Pillay S, Traynor AE, Rosen ST. Cyclic adenosine-3′,5′-monophosphate-mediated cytotoxicity in steroid sensitive and resistant myeloma. Clinical Cancer Research. 1997;**3**(10):1781-1787

[22] Kiefer J, Okret S, Jondal M, McConkey DJ. Functional glucocorticoid receptor expression is required for cAMP-mediated apoptosis in a human leukemic T cell line. Journal of Immunology. 1995;**155**(10):4525-4528

[23] Rangel-Aldao R, Allende O, Triana F, Piras R, Henriquez D, Piras M. Possible role of cAMP in the differentiation of *Trypanosoma cruzi*. Molecular and Biochemical Parasitology. 1987;**22**(1):39-43

[24] Aboagye-Kwarteng T, OK O-MY, Lonsdale-Eccles JD. Phosphorylation differences among proteins of bloodstream developmental stages of *Trypanosoma brucei brucei*. The Biochemical Journal. 1991;**275**(Pt 1):7-14

[25] Becker S, Jaffe CL. Effect of protein kinase inhibitors on the growth, morphology, and infectivity of *Leishmania promastigotes*. Parasitology Research. 1997;**83**(3):273-280

[26] de Castro SL, Meirelles MN, Oliveira MM. Trypanosoma cruzi: Adrenergic modulation of cyclic AMP role in proliferation and differentiation of amastigotes in vitro. Experimental Parasitology. 1987;**64**(3):368-375

[27] Hua SB, Wang CC. Differential accumulation of a protein kinase homolog in Trypanosoma brucei. Journal of Cellular Biochemistry. 1994;**54**(1):20-31

[28] Oliveira MM, Rocha ED, Rondinelli E, Arnholdt AV, Scharfstein J. Signal transduction in *Trypanosoma cruzi*: Opposite effects of adenylcyclase and phospholipase C systems in growth control. Molecular and Cellular Biochemistry. 1993;**124**(2):91-99

[29] Parsons M, Valentine M, Deans J, Schieven GL, Ledbetter JA. Distinct patterns of tyrosine phosphorylation during the life cycle of Trypanosoma brucei. Molecular and Biochemical Parasitology. 1991;**45**(2):241-248

[30] Parsons M, Valentine M, Carter V. Protein kinases in divergent eukaryotes: Identification of protein kinase activities regulated during trypanosome development. Proceedings of the National Academy of Sciences of the United States of America. 1993;**90**(7):2656-2660

[31] Sulakhe PV, Vo XT, Morris TE, Pato MD, Khandelwal RL. Protein phosphorylation in rat cardiac microsomes: Effects of inhibitors of protein kinase a and of phosphatases. Molecular and Cellular Biochemistry. 1997;**175**(1-2):109-115

[32] Rangel-Aldao R, Triana F, Comach G, Abate T, Fernández V, McMahon-Pratt D. Intracellular signaling transduction in the differentiation of *Trypanosoma cruzi*: Role of cAMP. Archivos de Biología y Medicina Experimentales. 1988;**21**(3-4):403-408

[33] Dremier S, Kopperud R, Doskeland SO, Dumont JE, Maenhaut C. Search for new cyclic AMP-binding proteins. FEBS Letters. 2003;**546**(1):103-107

[34] Hansen BD, Chiang PK, Perez-Arbelo J. Evidence for a membrane adenosine receptor in *Leishmania mexicana mexicana* (WR 227). Advances in Experimental Medicine and Biology. 1986;**195**(Pt B):547-551

[35] Hempel CM, Vincent P, Adams SR, Tsien RY, Selverston AI. Spatio-temporal dynamics of cyclic AMP signals in an intact neural circuitm. Nature. 1996;**384**(6605):166-169

[36] Wirth JJ, Kierszenbaum F. Inhibitory action of elevated levels of adenosine-3′:5′ cyclic monophosphate on phagocytosis: Effects on macrophage—*Trypanosoma cruzi* interaction. Journal of Immunology. 1982;**129**(6):2759-2762

[37] Mancini PE, Patton CL. Cyclic 3′,5′-adenosine monophosphate levels during the developmental cycle of *Trypanosoma brucei brucei* in the rat. Molecular and Biochemical Parasitology. 1981;**3**(1):19-31

[38] Gonzales-Perdomo M, Romero P, Goldenberg S. Cyclic AMP and adenylate cyclase activators stimulate *Trypanosoma cruzi* differentiation. Experimental Parasitology. 1988;**66**(2):205-212

[39] Sanchez MA, Zeoli D, Klamo EM, Kavanaugh MP, Landfear SM. A family of putative receptor-adenylate cyclases from *Leishmania donovani*. The Journal of Biological Chemistry. 1995;**270**(29):17551-17558

[40] Sacks DL, Perkins PV. Identification of an infective stage of *Leishmania promastigotes*. Science. 1984;**223**(4643):1417-1419

[41] Pays E, Tebabi P, Pays A, Coquelet H, Revelard P, Salmon D, et al. The genes and transcripts of an antigen gene expression site from *T. brucei*. Cell. 1989;**57**(5):835-845

[42] Ross DT, Raibaud A, Florent IC, Sather S, Gross MK, Storm DR, et al. The trypanosome VSG expression site encodes adenylate cyclase and a leucine-rich putative regulatory gene. The EMBO Journal. 1991;**10**(8):2047-2053

[43] Paindavoine P, Rolin S, Van Assel S, Geuskens M, Jauniaux JC, Dinsart C, et al. A gene from the variant surface glycoprotein expression site encodes one of several transmembrane adenylate cyclases located on the flagellum of *Trypanosoma brucei*. Molecular and Cellular Biology. 1992;**12**(3):1218-1225

[44] Biswas A, Bhattacharya A, Vij A, Das PK. Role of leishmanial acidocalcisomal pyrophosphatase in the cAMP homeostasis in phagolysosome conditions required for intra-macrophage survival. The International Journal of Biochemistry & Cell Biology. 2017;**86**:1-13

[45] Docampo R, Scott DA, Vercesi AE, Moreno SN. Intracellular Ca^{2+} storage in acidocalcisomes of *Trypanosoma cruzi*. The Biochemical Journal. 1995;**310** (Pt 3):1005-1012

[46] Seebeck T, Schaub R, Johner A. cAMP signalling in the kinetoplastid protozoa. Current Molecular Medicine. 2004;**4**(6):585-599

[47] Kudlacek O, Mitterauer T, Nanoff C, Hohenegger M, Tang WJ, Freissmuth M, et al. Inhibition of adenylyl and guanylyl cyclase isoforms by the antiviral drug foscarnet. The Journal of Biological Chemistry. 2001;**276**(5):3010-3016

[48] Conti M, Beavo J. Biochemistry and physiology of cyclic nucleotide phosphodiesterases: Essential components in cyclic nucleotide signaling. Annual Review of Biochemistry. 2007;**76**:481-511

[49] Bhattacharya A, Biswas A, Das PK. Role of a differentially expressed cAMP phosphodiesterase in regulating the induction of resistance against oxidative damage in *Leishmania donovani*. Free Radical Biology & Medicine. 2009;**47**(10):1494-1506

[50] Gettys TW, Vine AJ, Simonds MF, Corbin JD. Activation of the particulate low km phosphodiesterase of adipocytes by addition of cAMP-dependent protein kinase. The Journal of Biological Chemistry. 1988;**263**(21):10359-10363

[51] Vij A, Biswas A, Bhattacharya A, Das PK. A soluble phosphodiesterase in *Leishmania donovani* negatively regulates cAMP signaling by inhibiting protein kinase A through a two way process involving catalytic as well as non-catalytic sites. The International Journal of Biochemistry & Cell Biology. 2014;**57**:197-206

[52] Mochly-Rosen D. Localization of protein kinases by anchoring proteins: A theme in signal transduction. Science. 1995;**268**(5208):247-251

[53] Gimeno CJ, Ljungdahl PO, Styles CA, Fink GR. Unipolar cell divisions in the yeast S. cerevisiae lead to filamentous growth: Regulation by starvation and RAS. Cell. 1992;**68**(6):1077-1090

[54] Anjard C, Pinaud S, Kay RR, Reymond CD. Overexpression of Dd PK2 protein kinase causes rapid development and affects the intracellular cAMP pathway of *Dictyostelium* discoideum. Development. 1992;**115**(3):785-790

[55] Anjard C, Söderbom F, Loomis WF. Requirements for the adenylyl cyclases in the development of *Dictyostelium*. Development. 2001;**128**(18):3649-3654

[56] Hartmann A, Arroyo-Olarte RD, Imkeller K, Hegemann P, Lucius R, Gupta N. Optogenetic modulation of an adenylate cyclase in *Toxoplasma gondii* demonstrates a requirement of the parasite cAMP for host-cell invasion and stage differentiation. The Journal of Biological Chemistry. 2013;**288**(19):13705-13717

[57] Kurokawa H, Kato K, Iwanaga T, Sugi T, Sudo A, Kobayashi K, et al. Identification of *Toxoplasma gondii* cAMP dependent protein kinase and its role in the tachyzoite growth. PLoS One. 2011;**6**(7):e22492

[58] Mukhopadhyay NK, Saha AK, Lovelace JK, Da Silva R, Sacks DL, Glew RH. Comparison of the protein kinase and acid phosphatase activities of five species of Leishmania. The Journal of Protozoology. 1988;**35**(4):601-607

[59] Corbin JD, Soderling TR, Park CR. Regulation of adenosine 3′,5′-monophosphate-dependent protein kinase. I. Preliminary characterization of the adipose tissue enzyme in crude extracts. The Journal of Biological Chemistry. 1973;**248**(5):1813-1821

[60] Corbin JD, Keely SL. Characterization and regulation of heart adenosine 3′:5′-monophosphate-dependent protein kinase isozymes. The Journal of Biological Chemistry. 1977;**252**(3):910-918

[61] Taylor SS, Buechler JA, Yonemoto W. CAMP-dependent protein kinase: Framework for a diverse family of regulatory enzymes. Annual Review of Biochemistry. 1990;**59**:971-1005

[62] Spaulding SW. The ways in which hormones change cyclic adenosine 3′,5′-monophosphate-dependent protein kinase subunits, and how such changes affect cell behavior. Endocrine Reviews. 1993;**14**(5):632-650

[63] Banerjee C, Sarkar D. Isolation and characterization of a cyclic nucleotide-independent protein kinase from *Leishmania donovani*. Molecular and Biochemical Parasitology. 1992;**52**(2):195-205

[64] Malki-Feldman L, Jaffe CL. *Leishmania major*: Effect of protein kinase a and phosphodiesterase activity on infectivity and proliferation of promastigotes. Experimental Parasitology. 2009;**123**(1):39-44

[65] Bhattacharya A, Biswas A, Das PK. Identification of a protein kinase A regulatory subunit from Leishmania having importance in metacyclogenesis through induction of autophagy. Molecular Microbiology. 2012;**83**(3):548-564

[66] Genestra M, Cysne-Finkelstein L, Leon L. Protein kinase a activity is associated with metacyclogenesis in *Leishmania amazonensis*. Cell Biochemistry and Function. 2004;**22**(5):315-320

[67] Maurice DH, Ke H, Ahmad F, Wang Y, Chung J, Manganiello VC. Advances in targeting cyclic nucleotide phosphodiesterases. Nature Reviews. Drug Discovery. 2014;**13**(4):290-314

[68] Martinez A, Gil C. cAMP-specific phosphodiesterase inhibitors: Promising drugs for inflammatory and neurological diseases. Expert Opinion on Therapeutic Patents. 2014;**24**(12):1311-1321

[69] Wang P, Myers JG, Wu P, Cheewatrakoolpong B, Egan RW, Billah MM. Expression, purification, and characterization of human cAMP-specific phosphodiesterase (PDE4) subtypes A, B, C, and D. Biochemical and Biophysical Research Communications. 1997;**234**(2):320-324

[70] Drott J, Desire L, Drouin D, Pando M, Haun F. Etazolate improves performance in a foraging and homing task in aged rats. European Journal of Pharmacology. 2010;**634**(1-3):95-100

[71] Jindal A, Mahesh R, Bhatt S. Etazolate, a phosphodiesterase 4 inhibitor reverses chronic unpredictable mild stress-induced depression-like behavior and brain oxidative damage. Pharmacology, Biochemistry, and Behavior. 2013;**105**:63-70

[72] Yuasa K, Mi-Ichi F, Kobayashi T, Yamanouchi M, Kotera J, Kita K, et al. PfPDE1, a novel cGMP-specific phosphodiesterase from the human malaria parasite Plasmodium falciparum. The Biochemical Journal. 2005;**392**(Pt 1):221-229